Praise for the *Process Improvement Handbook:*

"I believe that this is a book that will end up with many yellow highlights in almost every chapter. You are wise to buy this book; you'll be even wiser if you apply its contents to the way your organization functions."

—H. James Harrington, Ph.D., prolific author of over 35 Process Improvement books, CEO Harrington Institute

"Tristan Boutros and Tim Purdie have delivered a true gift to all business process management professionals, and to the business community as a whole. Comprehensive in scope, deliberate in focus, deep in expert knowledge, and highly practical in its wisdom, *The Process Improvement Handbook* enables the reader to put it all together and take a more integrative approach to improving business performance. With so much already written on the topic of process improvement, the authors tackle the challenge of differentiation head on by defining a compelling business case for this work—identifying critical links to other enterprise management disciplines and defining key contributions to organizational agility and value innovation. If real knowledge is less a matter of knowing everything and more about knowing where to find everything, then this book is an indispensable resource for everyone seeking answers to performance challenges."

—Joseph A. Braidish,
Director of Consulting and Training,
Rummler-Brache Group

"Boutros and Purdie present a framework that is invaluable for any organization looking to improve its processes (and what organization isn't looking to do that). Concise, clear, and credible, this handbook is full of ideas and approaches that will benefit anyone working with Process Improvement."

—Mike Jacka, author of *Business Process Mapping: Improving Customer Satisfaction*, Second Edition

"Finally, we have book that takes Process Improvement from theory to templates. There are a lot of books that poke at pieces of Process Improvement structure, but this book finally takes us across the entire spectrum. It is invaluable in two ways: it is a valuable resource for identifying and understanding the role and impact of a process within an organization, and it is a valuable ongoing reference. I strongly recommend it to anyone who is searching for process improvement opportunities, no matter if a novice or advanced practitioner."

—Gerhard Plenert, author of *Strategic Continuous Process Improvement*

"The Process Improvement Handbook is an indispensable resource for achieving meaningful Process Improvement; business managers will have the power to help their organizations stay competitive and responsive to customers, as well as nimble and resilient. Readers will be miles ahead from where they would be if they tried to navigate the world of process management on their own."

—Ken Carraher,
CEO of iGrafx

"This approach works! This book builds on proven Process Improvement techniques refined over the past 30 years, but makes the case for a new standardized approach to BPI for organizations. It makes this case because it expertly weaves today's technology and capabilities, with the proven quality and process approaches of the past. This is a great book for those that want to lead or enact serious positive and sustainable change for their organization. The practical approach and templates included are invaluable and will enable a fast start!"

—Mike Morris,
Worldwide Sales Leader,
IBM Smarter Process

"The promise of enterprise software is to increase the effectiveness and competitiveness of organizations. And at the heart of that promise is process automation. Tim and Tristan have captured the fundamentals of Process Improvement in their new book."

—Mark J. Barrenechea,
President and Chief Executive Officer OpenText Corporation

"Often organizations approach business performance improvement with individual methodologies such as Six Sigma. While this approach can be effective, many organizations are faced with challenges that require a multifaceted approach to consistently deliver value to their customers. *The Process Improvement Handbook* is a valuable resource for all leaders looking to effectively and efficiently execute within their organizations."

—Brent Drever,
Chief Executive Officer, Acuity Institute

"A must read for any leader looking to impact bottom line growth and set the stage for operational excellence."

—Dayan M. Douse, Ph.D., best selling coauthor of the book *Nothing But Net*

"Enterprise level process improvement is often misunderstood and improperly executed by organizations. Finally, there is a book that shows you how to execute on continuous Process Improvement at all levels of your organization."

—Dave McCrory,
author of the book *Advanced Server Virtualization:
VMware and Microsoft Platforms in the Virtual Data Center*

"A unique and practical approach, in great detail, that gives organizations the tools and knowledge to take action quickly and successfully."

—Theo Mandel, Ph.D.,
author of *The Elements of User Interface Design*

"The challenge for all business leaders today is how to evolve business operations, organization structures and processes to survive in today's hyper-competitive world. *The Process Improvement Handbook* is the perfect companion for anyone responsible for implementing and managing processes for the Composable Enterprise."

—Jonathan Murray,
EVP & CTO at Warner Music Group

"*The Process Improvement Handbook* is an outstanding reference for anyone striving to build process excellence. Whether you are an experienced practitioner or a beginner just starting out, you'll find something worthy that you can learn and apply right away."

—Bernie Cardella,
SVP Supply Chain & Procurement at Arclin

"This is the 'go-to book' for practical understanding about how to build effective processes that focus on the human side of performance design."

—Lori Palmer,
Head of Organizational Effectiveness at Zurich Insurance

"This book has provided a service to all of us. The authors deliver tremendous value and I was blown away by the results the framework delivered!"

—Christopher J Barry,
President and General Manager,
The Leukemia and Lymphoma Society of Canada

"A powerful and comprehensive book that systematically breaks Process Improvement down and is easy to understand and apply. Senior executives to individual contributors using this book will produce valuable improvements in their organizations."

—Mike Inman,
Partner at TableForce, former Head of Procurement for
MGM Resorts International
and IAC/InterActiveCorp

"*The Process Improvement Handbook* has become a must read for all of our consultants. The techniques learned from this book have benefited not only our company, but our clients as well."

—William T. Sanders,
Founder and Managing Director, Summus Group

"*The Process Improvement Handbook* is the modern standard for Process Improvement. Not only will you learn the art of process improvement, it will fundamentally change the way you approach your daily work and make you more successful. This book is one that you keep within arms reach at your desk as it is filled with wisdom and helpful guidance you can use every day."

—Malti Raisinghani,
SVP Product Management &
Operations at Warner Music Group

"This is the Process Improvement book we've all been waiting for—the book that effectively combines Enterprise Architecture and disciplined Process Improvement. Not only does this book help us make sense of Process Improvement in this current world of iterative and incremental improvement, but it's an all-around good read!"

—Jason Gamet,
Vice President Operations, ePrize

"Finally, a practical, readable, intelligent process orientation model for business leaders who want to mature their understanding and, more importantly, take action."

—Roger Valade,
Vice President of Engineering at ProQuest

"In our ever-changing business landscape, understanding a business's core processes is essential to effect 'real' change and shifts in business models. Automation, mobility and social networking are driving business to rethink their core business processes. *The Process Improvement Handbook* is a must-have to engage the business on its continuous process journey."

—Robin Bienfait (retired) CIO BlackBerry and
AT&T Worldwide Network Services

The Process Improvement Handbook

A Blueprint for Managing Change and Increasing Organizational Performance

Tristan Boutros

Tim Purdie

Illustrations by Dustin Duffy

New York Chicago San Francisco
Athens London Madrid
Mexico City Milan New Delhi
Singapore Sydney Toronto

Cataloging-in-Publication Data is on file with the Library of Congress.

McGraw-Hill Education books are available at special quantity discounts to use as premiums and sales promotions, or for use in corporate training programs. To contact a representative please visit the Contact Us pages at www.mhprofessional.com.

The Process Improvement Handbook

1 2 3 4 5 6 7 8 9 0   DOC/DOC   1 2 0 9 8 7 6 5 4 3

ISBN 978-0-07-181766-0
MHID 0-07-181766-2

The pages within this book were printed on acid-free paper.

Sponsoring Editor
  Judy Bass

Acquisitions Coordinator
  Amy Stonebraker

Editorial Supervisor
  David E. Fogarty

Project Manager
  Sheena Uprety, Cenveo® Publisher Services

Copy Editor
  Mary Kay Kozyra

Proofreader
  Yamini Chadha, Cenveo Publisher Services

Indexer
  Robert Swanson

Production Supervisor
  Pamela A. Pelton

Composition
  Cenveo Publisher Services

Cover and Internal Illustrations
  Dustin Duffy

Art Director, Cover
  Jeff Weeks

Contents at a Glance

Contents

Part III Applying the Process Improvement Handbook

Forewords

"Anything that is of value is produced by a process."

Life begins and ends with a process. In between, we are involved in and subjected to millions of different processes. With processes being at the core of everything we encounter, it is strange that we have been so slow in realizing the importance of focusing our attention on improving these processes in our private and business life.

In the early 1900s, Frank Bunker Gilbreth, the father of Lean, was preaching the importance of process improvement. Gilbreth was a process method man. He felt there was one best way of doing things. He focused on the elimination of waste and flowcharted almost everything. In his 1921 book, *Process Charts*, he wrote "The Process Chart is a device for visualizing a process as a means of improving it. Every detail of a process is more or less affected by every other detail; therefore the entire process must be presented in such form that it can be visualized all at once before any changes are made in any of its subdivisions. In any subdivision of the process under examination, any changes made without due consideration of all the decisions and all the motions that precede and follow that subdivision will often be found unsuited to the ultimate plan of operation."

Gilbreth also applied this concept to the personal activities that went on within his home. For example, he insisted that all of his 12 children take a bath using the same process. This resulted in minimizing the cycle time and reducing the water and soap consumption. The process ensured that all 12 children took a bath that met the requirements of their customer (their mother and father). He applied the same approach to almost every one of the family's personal and household activities.

Looking at the process improvement activities in the latter half of the 20th century, World War II forced American industry to focus on manufacturing process improvement. At first, large quantities of defective finished goods were being delivered to our fighting forces. Since there was a high demand for military products and a very limited amount of natural resources available to produce them, something had to be done. As a result, a major focus was applied to our manufacturing processes. Routings that controlled the manufacturing processes were refined. Sampling plans were put in place and processes were certified.

In 1961, Phil Crosby launched the first Zero Defects program at the Martin Company's Orlando, Florida facility, in an effort to drive down the number of defects in the Pershing missile to one half of the acceptable quality level in half a year's time. The focus here was on improving worker performance, thereby eliminating or minimizing the inspection effort in the process and the following process rework time. The Zero Defects program flourished in the 60s and early 70s. For those of you that think that Zero Defects is too stringent a target, you can compromise and accept Six Sigma as an acceptable performance standard.

Up to the mid-1960s, most, if not all of industrial and service organizations' process improvement efforts focused on the product delivery processes. In the mid-1960s, IBM's San Jose plant initiated a program called "White-Collar Quality Costs." This initiative focused on flowcharting the support processes and eliminating waste from them. In 1980, IBM undertook a worldwide initiative to focus on improving the quality of IBM's products and services internal and external to the organization. The concept of focusing attention on flowcharting and eliminating waste in the support processes was included in this initiative. It wasn't long before enough data was collected to prove that there was a large amount of gold to be mined by improving the performance of the support processes. As a result, in 1983, a worldwide initiative focusing on Business Process Improvement was begun. Shortly after that, a directive came out from the President of IBM essentially stating that he expected all of the support processes to be functioning at a performance level that was equal to or better than IBM's manufacturing processes. Process owners were assigned to all the major cross-functional processes and a five-step maturity grid was established and used to evaluate and set targets for the individual processes. A directive requiring process compatibility across plants internationally was also released in the early 1980s at IBM.

By the end of the 1980s, Business Process Improvement approaches were beginning to be applied around the world. This focus on Business Process Improvement escalated in the early 1990s with the release of two books—mine, *Business Process Improvement* (McGraw-Hill 1991) and *Reengineering the Corporation* by Michael Hammer and James Champy (Nicholas Brealey Publishing 1993).

Over time, normal flowcharting gave way to swim lane flowcharting, which gave way to rainbow flowcharting, which led to critical path analysis.

A series of other mapping approaches like graphic flowcharting, knowledge flow maps, and information flow maps have been developed to help the process improvement teams design more efficient, effective and adaptable processes. Today simulation modeling is a leading-edge approach to focusing on the elimination of waste from our processes. All these changes have been made to increase the focus on the waste that occurs throughout our processes. Business Process Improvement, Process Reengineering, Total Quality Management, Six Sigma, or Lean Six Sigma approaches all were designed to improve the organization's processes, but they only provide temporary improvements in the organization's performance. In Michael Hammer and James Champy's 1984 book, *Reengineering the Corporation* they wrote "More than 50% of all the massive restructuring and reengineering projects fail." Numerous other magazine articles and presentations have reported similar results from the other process and performance improvement methodologies.

Today it is widely accepted that the key to business success is process breakthrough followed by continuous improvement of all the organization's processes. Many short-sighted organizations have limited this thinking to their customer-related processes not recognizing that any process worthwhile doing is worthwhile improving.

This book provides a step-by-step approach to bringing about substantial improvement within your processes that will result in significant bottom-line performance improvement. It highlights the importance of integrating the key process principles into the culture within the organization and provides the reader with tools and techniques that can guide them in making this important cultural change. *The Process Improvement Handbook* distinguishes itself from other books related to process improvement by providing the reader with significant insight into how to bring about this required cultural change. Almost all the other process improvement books have emphasized the importance of cultural change in support of their methodology without providing the reader with direction in how to accomplish the desired cultural change. The major cause of past performance improvement methodology failures was the result of short-term improvement that was not sustained and rarely reflected in the organization's bottom line or long-term performance. This is a rare book that provides guidance in improving your processes as well as preparing the organization to accept a new operating philosophy.

In Chapter 3 a comprehensive process maturity grid is presented. This provides an effective means of directing your process improvement evolution that will effectively support the necessary cultural change.

I believe that this is a book that will end up with many yellow highlights in almost every chapter. You are wise to buy this book; you'll be even wiser if you apply its contents to the way your organization functions.

Dr. H. James Harrington
CEO, Harrington Institute

Disciplined process execution is the backbone of all world-class organizations. The increasingly complexity of globally distributed production and supply chains, changing consumer preferences and ever shorter product cycles makes effective and repeatable process orchestration an absolute requirement for successful business execution.

Business processes can no longer be viewed as static: once implemented, never to be changed. Today's competitive markets demand dynamic response and flexibility enabled by business models and processes that adapt to continuous change.

In the modern enterprise, the synergy between business processes and enabling information technology services is inescapable. The adoption of a flexible 'service architecture'-based approach to designing business processes—a Process Oriented Architecture—mirrors the best practices now being adopted in highly scalable and flexible IT systems. Utilizing a common architectural approach between Process and IT system design is the key to building truly adaptive business operations.

The challenge for all business leaders today is how to evolve business operations, organization structures and processes to survive in today's hyper-competitive world. *The Process Improvement Handbook* is the perfect companion for anyone responsible for implementing and managing processes.

Jonathan Murray
Executive Vice President &
Chief Technology Officer
Warner Music Group

Preface

Today's business environment requires organizations to be fast, flexible, and fluid. As a result, corporate cultures and subsequent operational practices have had to adapt and evolve to meet ever-changing market requirements. Executives know their business is dependent on constantly adapting to these shifts in the marketplace, and adjusting business processes has become a primary means of sustaining industry competitiveness. Adopting a process-oriented perspective can assist organizations with delivering attractive products and services to customers, add efficiency to the way in which they operate, and in many cases facilitate their survival and prosperity in the face of competition. Whether an enterprise is focused on efficiency gains, innovation, growth, cost reductions, or customer satisfaction, Process Improvement has become an essential ingredient for business sucess.

Unfortunately, wide adoption of Process Improvement as a discipline has inspired an increased development of models and methods, and this proliferation has led to conflicts in goals and techniques, wasted investments in competing training programs, and overall confusion about which of the various models best applies to an organization's specific needs. As a result, developing flexible processes and finding trained practitioners to facilitate improvement efforts has become increasingly difficult. Organizations have found themselves struggling to attract, develop, motivate, organize, and retain talented people who possess the training and foundations of Process Improvement needed to deliver a unified and winning outcome.

Over the years, there have been many texts that focus on individual Process Improvement methodologies, but very few that help professionals understand Process Improvement as an end-to-end enterprise construct, or the issues involved in designing effective Process Improvement structures, organizations, and cultures. Moreover, we have observed several other recurrent problems that have led to organizational frustration and unsuccessful or undesirable Process Improvement outcomes. These include the following:

- Process Improvement efforts are often carried out without an enterprise perspective in mind, leading to situations where the wrong problem or only a fragment of the real problem has been resolved.

- Process Improvement efforts often overlook how process changes will affect the employees who will interact with them, leading to stakeholder resistance to future improvement engagements.

- Process models and designs are often executed without following a set of proper architecture principles, leading to opposing or redundant implementations.

- Process Improvement Managers are often confused about which Process Improvement method to use and the scenarios in which certain methods are most appropriate.

- Many Process Improvement projects are very complex and have multiple and changing objectives. In addition, current Process Improvement methods do not incorporate agility, making the development and application of improvements more challenging.

- Process Improvement vocabulary and concepts are often presented in confusing, unintelligible, or unfriendly ways to stakeholders, leading to business fatigue and frustration.

- Organizational Process documentation is often stored in disparate systems or locations with no centralized view that is easy to explore or navigate.

- Process Improvement organizations lack a comprehensive framework where enterprise architecture principles and process modeling and improvement concepts are strategically aligned and have standardized definitions.

- Process Improvement lacks a common set of roles or titles to describe practitioners and professionals similar to other disciplines, such as Project Management or Business Analysis.

- Process Improvement lacks a comprehensive standard or body of knowledge that defines the necessary terminology and concepts in an unambiguous way.

In a business environment that demands faster responses, better service, and increased agility, organizations must ensure that Process Improvement is an enabler of organizational change and not a hindrance. With the huge growth in spending on Process Improvement by enterprises and the strong evidence that significant investment in this domain can lead to cost savings and better business decision-making, the time has come to make the Process Improvement discipline more professional. *The Process Improvement Handbook* serves to establish the foundation and instruction needed to create, maintain, measure, and flex processes while equipping practitioners with the necessary knowledge to create consistent and successful outcomes. This handbook is intended to provide a framework and a set of tools and principles that can substantively address the problems that often burden the Process Improvement profession.

TRISTAN BOUTROS
TIM PURDIE

About the Authors

Tristan Boutros is vice president and chief process officer for one of the world's largest media and entertainment firms in New York City where he oversees program management, process improvement, and enterprise architecture. He holds more than 10 professional designations including his Lean Six Sigma Black Belt, Project Management Professional, Certified Scrum Professional, and Master Project Manager certification. He has more than 10 years of business, technology, and management consulting experience at companies including Pernod Ricard, IAC, DTE Energy, BlackBerry and Warner Music Group and has been a successful entrepreneur. A skilled facilitator and change agent, he specializes in delivering rapid business value using numerous techniques and methods including the Process-Oriented Architecture and Process Ecosystem approaches.

Tim Purdie is vice president of global information technology for a high-tech company in Waterloo, Ontario, Canada. He is a Lean Six Sigma Black Belt, Master Project Manager, certified Scrum Master, and certified Rummler–Brache Process Improvement consultant with more than 25 years of high-tech information technology, management consulting, with advanced process implementation experience. A skilled public speaker, he specializes in turnaround ventures and delivering rapid business value using the Process-Oriented architecture and Process Ecosystem approach. He has experience at companies including Sandvine, BlackBerry, IAC/InterActiveCorp, General Motors, DTE Energy, Western International Communications, and OpenText.

The authors have worked together for more than six years, collaborating and building business success. They have formalized their process improvement experience into *The Process Improvement Handbook*.

Acknowledgments

We acknowledge the many individuals and groups who have made this book possible.

Leadership in Process Improvement

First, we thank those who have pioneered and introduced the many methods that served as inspiration for writing this text. Without your thought leadership, knowledge sharing, and dedication to the industry, this text would not be possible. *The Process Improvement Handbook* draws on the topics of Enterprise Architecture, Project Management, Process Management, Enterprise Modeling, Lean Six Sigma, Continuous Improvement, Capability Maturity Model, Decision Modeling, Business Intelligence, Agile, and other bodies of knowledge as well as author experience to build a comprehensive framework for Process Improvement professionals. In particular, we thank Dr. H. James Harrington for his support and encouragement as we completed this text.

Professional Colleagues

Second, we thank our professional colleagues who helped to prove the concepts of *The Process Improvement Handbook* in the real world and who applied the Process-Oriented Architecture and Process Ecosystem approaches in their workplaces. These colleagues also heavily influenced our own education and encouraged us to share our lessons learned with you: Christopher Barry, Roger Valade, Jason Gamet, Jonathan Murray and Joseph Braidish. In particular, we thank Tracy Fischer, Michael Lukacko, and the leadership team at Blackberry for creating the unique opportunity to build an improvement organization from the ground up and supporting us as we deployed and demonstrated how this framework could be applied quickly and successfully.

We also thank McGraw-Hill Professional and our editorial team, Judy Bass and Amy Stonebraker, who provided encouragement and professional guidance to keep us on track. Our illustrator and idea man, Dustin Duffy, is also to be commended for working under pressure to bring to life the concepts and ideas that abound in the text. Dustin's tireless efforts help tell the story the way we always wanted it told.

Employers and Staff

Third, we thank our staff at the various companies where we have been employed who provided beyond-the-call-of-duty support to us as we created, piloted, and deployed this framework. Special thanks go to Hunain Mahmood, Laura Duffy, Dawn Dixon, Shawn McIntyre, Dane Bannister, Joanna Henderson, Abdul Zalmay, Angela Maloney, Jason Hinchliff, Corey Zanderson, and Trent Webber as well as the numerous other professional contacts at Entertainment Publications, Sandvine, Warner Music Group, and BlackBerry.

Families

Finally, we offer our most special thanks to our families who enthusiastically encouraged and supported us throughout this effort. *The Boutros Family*: Jennifer, Preston, Jacques and Shannon, Kalen, Bernie and Fernande, Angie, Everal and Donna, Michael and Julia and *The Purdie Family*: Lynda, Drew, and Riley.

The Process Improvement Handbook

PART I

Introduction

CHAPTER 1
Introduction

This chapter introduces *The Process Improvement Handbook* as a body of knowledge for Process Improvement professionals. Process Improvement information and its application come from a variety of sources, have different levels of aggregation, incorporate different assumptions about the topic, and manifest other differences that need reconciliation before practitioners can effectively use the concepts for improvement efforts within an organization. This chapter lays the groundwork for readers as they navigate the handbook, looks at some of the bases upon which the handbook was created, outlines the reasoning behind aggregating the content into a formal body of knowledge, and introduces the core values to which this guide was created. Chapter 1 is organized around the following topics:

- *Overview:* What is *The Process Improvement Handbook* and why was it created?

- *Purpose:* What are the key benefits to adopting and using the handbook?

- *Audience:* Who is the core audience for the handbook and who would benefit from reading it?

- *Navigating the Handbook:* What are the different components that make up the handbook?

- *Relationships to Other Management Disciplines:* What other disciplines does the handbook draw from and how do they complement one another?

- *Process Improvement Manifesto:* What are the core values that inspired the handbook and serve to drive all Process Improvement efforts?

- *Adapting the Handbook to Your Needs:* Is *The Process Improvement Handbook* applicable to all organizations?

- *Contacting Us:* What is the publisher's contact information? How can readers connect with the authors directly? How is the handbook's Resource Center accessed?

Overview of *The Process Improvement Handbook*

A handbook is a collection of instructions that are intended to provide ready reference in a formal text. These instructions describe tools, techniques, methods, processes, and practices. This handbook delivers an introduction to the key concepts in the Process Improvement field and summarizes the various techniques and principles considered good practice within the industry. In addition, it is a framework that describes the Process Improvement knowledge areas that must be considered in order to properly deliver value through Process Improvement efforts to an organization.

Implementing a high level of process maturity is a vital component toward unlocking performance improvement potential in organizations. Nonetheless, businesses have been forced to spend precious time hunting down practical frameworks or have had to create their own in trial-and-error fashion, eroding the core value of realizing gains quickly in order to better compete in the marketplace. As a result, businesses have struggled to implement Process Improvement frameworks that drive adoption and consistency, enable leaner operations, and create repeatable methods for building organizational maturity. Companies that attempt to build organizational maturity without a formal Process Improvement framework often spend significant time and money with less-than-ideal results, overtax key employees to create extensive documentation, execute process improvements without providing adequate assistance, hire expensive consultants who build processes that can be used only once, are difficult to grow and maintain, and do not create a long-term foundation for success. *The Process Improvement Handbook* has been written to address these common difficulties and provides an industry standard for the practice of Process Improvement.

The Purpose

The Process Improvement Handbook provides thinking tools for anyone interested in improving their operating environment. It is a toolkit for translating widely accepted principles and methods into effective practices that can be adapted to fit readers' unique environments. Generic Process Improvement techniques have been adopted by a significant number of Fortune 500 companies as well as many small and mid sized organizations. Its application in for-profit and nonprofit organizations, both flourishing and struggling, as well as large and small corporations is a reflection of its success as a discipline. While many texts focus on specific Process Improvement methods, and driving individual Process Improvement projects in isolation, this text looks beyond traditional approaches to provide an all-encompassing guide. The result is process self-sufficiency, simple extensibility, greater sustainability, higher quality, and overall speed to value realization that improves competitive advantage ranking, no matter the industry.

This handbook serves as a baseline that practitioners can use to discuss the work they do and ensure that they have the skills needed to effectively perform the function. Also, it defines the skills and competencies that people who work with and employ Process Improvement professionals should expect a practitioner to demonstrate.

This handbook provides the following key benefits:

- A comprehensive body of knowledge that outlines the methods, tools, and roadmaps used to create sustainable Process Improvement efforts and ensure consistent application

- A self-sufficient reference guide that all employees can easily use or self-train with

- A common vocabulary within the Process Improvement profession for discussing, writing, and applying Process Improvement concepts

- An easy-to-understand foundation for process maturity capability in any company

- An industry-leading architecture approach for building organizational maturity

- A framework that structures agile process adoption for rapid growth

- A robust tool for educating and training organizations and professionals

- Templates and real-life examples for implementing Process Management and Improvement concepts

- A roadmap for implementing a robust Process Improvement Framework

The Audience

The Process Improvement Handbook was created as a reference tool for anyone involved in creating, managing, operating, or improving products and processes to attain superior results. In particular, it was created to support the various specialists who carry out Process Improvement tasks, such as Continuous Improvement Professionals, Process Improvement Managers, Process Analysts, Enterprise Solution Architects, and Consultants. These specialists base their core profession on creating, maintaining, and teaching others how to become process mature and drive performance improvement. The text is intended to provide comprehensive information about the professional standard of work expected in improvement efforts and to establish practical guidance for those interested in taking on Process Improvement in their organization.

This text also serves as a playbook for all levels in the organizational hierarchy including first-level supervisors through C-level executives, as well as individual contributors across the various functions and lines of business engaged in producing a company's products or services. This may include Information Technology Professionals, Supply Chain Professionals, Operations Managers, Customer Service Specialists, Marketing Managers, Software Architects, Application Developers or Designers, Project and Program Managers, Security Professionals, as well as Finance and Legal Managers. The text is applicable to anyone who needs to understand the end-to-end experience of their employees and customers, those with a desire to improve efficiency and standardization within their role, and those who wish to improve their organization by achieving higher process maturity levels.

This handbook also serves as an aid to stakeholders who may not be directly involved with leading, improving, or operating processes but have a need to know about the outcomes produced and the associated effects. These stakeholders include customers, suppliers, users, the public, the media, and government regulators.

There is no prerequisite training or skill required to gain value from this text. Novice readers will gain foundational instruction that will enable them to make effective use of the material, and intermediate and advanced individuals will benefit from renewed approaches to the core tenants of process delivery, with a focus on process architectures in a rapid value-realization environment. This book is for individuals who wish to improve the way in which they conduct day-to-day operations as well as implement large-scale change within their organizations.

Navigating the Handbook

The Process Improvement Handbook is divided into three core modules: an introduction to the Process Improvement Body of Knowledge, Process Management Knowledge Areas, and practical information for Applying the Process Improvement Body of Knowledge. Part I builds understanding of Process and Process Improvement. Part II provides an overarching Process Architecture; identifies requisite skills; and offers tools, templates, and instructions for its practical application. Part III contains the appendices, which provide tools to better navigate this text and apply what is learned in real-world situations. Each part is described in detail below.

Part I—Introduction

Chapter 1, Introduction: The introductory chapter presents a basis and purpose for the standard. It provides an overview of the handbook and discusses its relationship to other management disciplines and how to adapt it to an organization's needs. It also describes its position as a Body of Knowledge and competitive advantage for organizations, departments, employees, and students.

Part II—The Process Improvement Knowledge Areas

Chapter 2, The Process Improvement Context: This chapter describes the fundamentals involved in Process Improvement and the key terms used to drive these efforts. It addresses the common pitfalls of traditional approaches to process adoption and outlines the role of a Process Improvement Manager. This chapter discusses the differences between a functional and a nonfunctional process management structure and the value proposition of a Process Improvement Organization.

Chapter 3, Process Maturity: This chapter outlines a set of structured levels that describe how well an organization's behaviors, practices, and processes can reliably and sustainably produce required outcomes. It describes an evolutionary improvement path that guides organizations as they move from immature, inconsistent business activities to mature, disciplined processes. This chapter enables management to evaluate where their organization stands relative to other organizations in their industry.

Chapter 4, Process-Oriented Architecture: This chapter defines a philosophical approach to process interaction and management. It contains a set of architectural elements that are used to sustain business performance, deliver a consistent experience, and enable continuous improvement in enterprises. The chapter provides a new industry architecture approach to building process maturity and sustainability. In addition, it also describes both a framework and a method for applying process-oriented architecture governance in an organization and making it work regardless of an organization's idiosyncrasies.

Chapter 5, Creating a Process Ecosystem: This chapter describes the management of an enterprise as an ecosystem in which all processes are interconnected and driving toward business success. It provides an overview of the elements needed to orchestrate and choreograph an end-to-end business process ecosystem and offers a basic model for practitioners to enhance or develop as needed.

Chapter 6, Managing Process Improvements: This chapter outlines the Process Improvement framework and provides an overview of the various environmental factors that can affect Process Improvement efforts. It discusses the various Process Improvement methods and includes an overview of an organizational structure that can influence an organization's culture and the way it manages continuous improvement.

Chapter 7, The Process Improvement Organization: This chapter describes the Process Improvement Organization, lists the professional services offered by the department, and defines a baseline governance process for

Process Improvement activities. It helps establish the roles and responsibilities needed to help Process Improvement team members understand their accountability. This chapter incorporates best practices, controls, and approaches to ensure effective process execution.

Chapter 8, Process Improvement Aptitudes: This chapter identifies the skills, underlying competencies, and capabilities needed to drive results within a Process-oriented Enterprise. It focuses on the individual inventory of aptitudes needed and how to recognize and build them within an organization.

Part III—Applying the Process Improvement Body of Knowledge

Chapter 9, Case Examples: In this chapter, examples and guidance on how to avoid common pitfalls associated with Process Improvement implementation are provided. The chapter is built around real-life events that point out the value of positive, consistent, and reliable service experiences. It provides examples of how improvements can be made using a Process Improvement framework such as *The Process Improvement Handbook*.

Chapter 10, Process Improvement Templates and Instructions: Templates and instructions that can be used by Process Improvement Organizations to build and deliver consistent and sustainable processes are included in this chapter. The chapter outlines a variety of templates and links them to the various phases of the Process Improvement lifecycle.

Chapter 11, The Process Improvement Handbook Summary: This chapter reviews the foundation, framework, tools, and principles presented in the core chapters. The culmination of learning is provided as a competitive advantage for organizations, departments, employees, and students. This chapter summarizes the body of knowledge and enables the reader to tackle the challenge of process work with a new degree of confidence and knowledge.

Relationship to Other Management Disciplines

Most tools and techniques needed to manage processes and subsequent improvements are unique or nearly unique to Process Improvement. However, Process Improvement does overlap with other general management fields. General management encompasses planning, organizing, and monitoring the operations of an ongoing enterprise. The following general management areas may overlap with those presented in *The Process Improvement Handbook* and are closely linked to Process Improvement efforts:

- Sales, Operations, and Supply Chain Management
- Organizational Behavior and Human Resource Management
- Information Technology and Information Systems
- New Product Development, Marketing, and Research and Development

- Vendor Management, Contracts, and Legal
- Portfolio Management and Program/Project Management
- Finance and Accounting
- Customer Service and Support

Process Improvement Manifesto

This handbook is built on a set of several interrelated core values, which are referred to as the Process Improvement Manifesto (Figure 1-1). These values are embedded beliefs and behaviors found in high-performing organizations and serve as the basis for this text. They are the foundation for integrating key performance and operational requirements within a result-oriented framework. In addition, they create a basis for action and feedback that readers and professionals can use to deploy the concepts presented.

Process Improvement Manifesto

Agility	Process Improvement values agile and iterative improvement.	**Discipline**	Process Improvement values organizational discipline and maturity.
Quality	Process Improvement values quality in all aspects of delivery, from process creation to retirement, including process, people and technology changes.	**Enterprise Perspective**	Process Improvement values the consideration of what is best for the organization as a whole rather than specific departments, focus areas, geographies, or individuals when making decisions and conducting improvement work.
Leadership	Process Improvement values leadership that is proactive and open to ideas for improving all aspects of an organization.	**Service Orientation**	Process Improvement values the notion that Process Improvement organizations provide a service to companies, departments, sponsors, individuals, the community, the consumer, and the profession.
Communication	Process Improvement values open communication and participative decision-making throughout the entire Process Improvement lifecycle.	**Continuous Learning**	Process Improvement values training and educating those involved in Process Improvement efforts.
Respect	Process Improvement values collegial working relationships throughout Process Improvement efforts.	**Human Centered Design**	Process Improvement values the consideration of what is best for customers of a process (operators and end-consumers) when designing and implementing process solutions and improvements.

FIGURE 1-1 The Process Improvement Manifesto

Agility

The Process Improvement Handbook values agile and iterative improvement. Success in today's ever-changing market demands agility, a capacity for rapid change, and flexibility. Because change is inevitable, organizations that wish to continually improve must be able to nimbly adjust to and take advantage of emerging opportunities. This involves focusing on flexible work and planning practices that are tailored to incremental improvement.

Quality

The Process Improvement Handbook values quality in all aspects of Process Improvement, from process creation to retirement, including process, people, and technology changes. Organizations that understand and focus attention on all facets of quality from the beginning of transformation initiatives to the end experience superior results. This involves the acts of process monitoring and control, ensuring a proper definition of quality is instituted, and upholding the highest standards in whatever tasks are performed.

Leadership

The Process Improvement Handbook values leadership that is proactive and open to ideas for improving all aspects of an organization. Leaders who communicate and inspire a clear and compelling vision for the future have teams that are more engaged and open to improvement opportunities. This involves creating an environment that inspires and enables everyone to contribute to the vision, be innovative, and achieve things of extraordinary value.

Communication

The Process Improvement Handbook values transparency and open communication, along with participative decision-making throughout the entire Process Improvement lifecycle. An organization that recognizes that everyone has a point of view and should have the opportunity to voice opinions, ideas, and experiences is generally more innovative in their improvement designs. This involves providing various mechanisms for people to participate in decisions that affect them when at all possible and ensuring that all ideas and suggestions offered are valued and considered for improving the enterprise.

Respect

The Process Improvement Handbook values collegial working relationships throughout Process Improvement efforts. An organization's success depends increasingly on an engaged workforce that has a safe, trusting, and cooperative work environment. Successful organizations capitalize on the diverse backgrounds, knowledge, skills, creativity, and motivation of their workforce and partners. Valuing the workforce means committing to their engagement, satisfaction, development, and well-being. This involves implementing more flexible, high-performance work practices that are tailored to varying workplace and home life needs and ensuring that project teams are properly formed and managed.

Discipline

The Process Improvement Handbook values organizational discipline and maturity. Companies with high organizational discipline that perform business processes in a standard, repetitive fashion are more competitive and usually leaders in their markets. Ensuring a disciplined approach to all Process Improvement activities helps ensure that thorough and robust solutions are implemented. This involves deploying standardized templates, performance expectations, auditing and inspection criteria, and shared enterprise systems to help achieve standardization across all areas of the business.

Enterprise Perspective

The Process Improvement Handbook values the consideration of what is best for the organization as a whole, rather than specific departments, focus areas, geographies, or individuals, when making decisions and conducting day-to-day work. Ensuring process improvements meet not only the needs of those involved with the activities in question, but also the larger enterprise ensures time and money are not wasted deploying and redeploying solutions. This involves guaranteeing that no duplication has occurred, previous improvements and lessons have been leveraged, and proper change management is in place.

Service Orientation

The Process Improvement Handbook values the notion that Process Improvement practitioners provide a service to companies, departments, sponsors, employees, the community, the consumer, and the profession. This involves doing what is right for the customer in question and endlessly providing expertise for their benefit.

Continuous Learning

The Process Improvement Handbook values training and educating those involved in Process Improvement efforts. The primary objective of training is to provide all personnel, suppliers, and customers with the skills needed to effectively perform quality process activities and to build this concept directly into an organization's operations. This practice makes continuous learning within the organization possible and promotes improvement and process-oriented thinking. Training can be proactive or just-in-time in its approach, depending on the needs and financial means of the department or organization.

Human-Centered Design

The Process Improvement Handbook values the consideration of what is best for customers of a process (operators and end consumers) when designing and implementing process solutions and improvements. Ensuring processes are user friendly for those executing its activities helps maintain positive morale. This involves ensuring the needs, wants, and limitations of process operators and end users are given extensive attention and consideration at each stage of process design and improvement.

Organizations that embrace these core values are capable of

- Quickly adapting to changing requirements or market factors
- Significantly reducing the risk associated with continuous improvements
- Accelerating the delivery of business value to customers
- Ensuring that value is continually being maximized throughout the Continuous Improvement process
- Meeting customer requirements faster and more efficiently
- Building innovation and best practices that help reach new maturity levels
- Discovering hidden knowledge and expertise within their workforce
- Improving performance and motivation across all areas of the business

Adapting the Handbook to Your Needs

This handbook is designed to be a critical resource for both students and practicing professionals in the Process Improvement field. By becoming proficient in the topics described, individuals will build a solid foundation on which comprehensive, organization-specific implementations can be established. However, it is our belief that corporations and organizational entities have unique requirements that are dependent on special environmental, cultural, customer, and market demands. As a result, readers will be required to apply the knowledge derived from this text by adapting it to fit various situations. Upon customization, the standards and principles contained herein provide a significant competitive differentiator. This knowledge will serve as the foundation for truly efficient organizations that are mature and agile, enabling them to make competitive moves much easier and faster without breaking moral or organizational structures. Some information provided in *The Process Improvement Handbook* can be complex and often focuses on key concepts and terminology. Rather than providing specific guidelines for action, this handbook provides basic knowledge on Process Improvement concepts and tools.

Contacting Us

Please address comments and questions concerning *The Process Improvement Handbook* and its distribution to the publisher:

McGraw-Hill Professional
1221 Avenue of the Americas, 45th Floor
New York, NY 10020

For more information about McGraw-Hill books, conferences, and resource centers, please visit their website at: www.mhprofessional.com.

We have a web page where we list errata, examples, templates, and additional information. You can access this page at: www.mhprofessional.com/pihandbook.

To comment on the material featured in this text, suggest future additions, ask technical questions, contact us about partnership, consulting, or training opportunities or to ask questions, make suggestions, or provide feedback, please reach out and connect via our professional profiles:

Tristan Boutros: linkedin.com/in/tristanboutros/or
Tim Purdie: ca.linkedin.com/in/timpurdie.
Follow Tristan Boutros on Twitter @TristanBoutros

Chapter Summary

In this chapter, we introduced *The Process Improvement Handbook* as a Body of Knowledge for Process Improvement professionals. This chapter outlined how to navigate the handbook, described the critical components that make up the text, and introduced the core values that make up the Process Improvement Manifesto. We also learned the following:

- Implementation of a high level of process maturity is a vital component toward unlocking an organization's performance improvement potential.

- *The Process Improvement Handbook* includes thinking tools for anyone interested in improving their operating environment.

- The handbook provides readers with a comprehensive guide that outlines the methods, tools, and roadmaps needed to ensure consistent application.

- *The Process Improvement Handbook* overlaps with several general management areas including Sales, Operations and Supply Chain Management, Organizational Behavior, and Human Resource Management, as well as Information Technology and Information Systems.

- There is no prerequisite training or skill required to gain value from this text. It is designed to be a critical resource for both students and practicing professionals at any level within an organization.
- Ten core values including agility, respect, and continuous learning make up the Process Improvement Manifesto.

Chapter Preview

Chapter 2 discusses the fundamental terminology used throughout Process Improvement efforts. In addition, it discusses the advantages of structuring an organization by process rather than by function and also discusses the role(s) a Process Improvement Manager plays in Process Improvement projects.

PART II

The Process Improvement Knowledge Areas

CHAPTER 2

The Process Improvement Context

This chapter covers the basic concepts of Process Improvement and introduces the fundamental conventions and learning blocks needed to discuss Process Improvement in practical terms. One of the most important components to successful Process Improvement efforts is ensuring that all participants are trained in the discipline, understand critical linkages to other areas, and have a common understanding of basic terminology. Ensuring this level of understanding occurs ahead of all other activities, helps reduce communication gaps, speeds up execution efforts, and fosters more collective thinking because learning roadblocks are removed early on. This chapter defines several key terms and provides the foundation for anyone looking to improve the state of business activities within their organization through Process Improvement.

This chapter focuses on the following knowledge areas:

- *Process defined:* What is a process? What is a subprocess? What are the different types of processes?

- *Relationships among Policies, Processes, and Procedures:* What is the difference between a Policy, a Process, and a Procedure?

- *Process Improvement defined:* What is Process Improvement? What is the value of conducting Process Improvement efforts?

- *Process Improvement Organization:* What is a Process Improvement Organization?

- *Process Improvement and Operations Management:* How does Process Improvement differ from day-to-day Operations Management?

- *Role of a Process Improvement Manager:* What is a Process Improvement Manager and what are his or her primary responsibilities?

- *Process Orientation and Organizational Structure:* What is the difference between organizing your business by function versus process, and why does it matter?

- *Leadership and Process Improvement:* What role does leadership play in Process Improvement efforts?

Process Defined

Every organization is made up of a series of interacting activities that are carried out in order to achieve intended results through a systematic and efficient allotment and usage of resources. A *Process* is a set of activities that use these resources (people, systems, tools) to transform inputs into value-added outputs. It is a sequential set of related tasks or subprocesses performed to achieve a particular business objective or to produce a specific product, service, good, or piece of information. A process is often thought of in the context of a workflow, user instructions, or the steps required to produce something of value to an organization's customers. Customers are both internal and external and include each person involved in the process as well as the entity that receives the final output of the process. Examples of processes include processing orders, invoicing, shipping products, updating employee information, and producing reports. Processes exist at all levels of an organization and include actions that the customer is able to see as well as actions that are invisible to the customer.

In general, the various activities of a business process can be performed in one of two ways: (1) manually by a business operator or (2) by means of interaction or automation with a technical system or application. Technical activities are typically a collection of substeps or instructions handled and processed by a particular system involved in the execution of activities contained within a process. In most cases, manual and system-based tasks may be sequenced in any order, and the data and information being handled through a process passes through a series of both manual and computer tasks. The outcome of a well-designed process is increased effectiveness (value for the customer) and increased efficiency (decreased costs for the company).

The following list outlines the general characteristics of a process:

- *Defined:* Processes have defined customers, boundaries, inputs, and outputs.
- *Sequential:* Processes consist of activities that are ordered to achieve specific results or outcomes.
- *Valuable:* Processes must add value to an organization and its customers.
- *Customer centric:* Processes are designed with an organization's operators and customers in mind.
- *Embedded:* A process exists within an organizational or departmental structure.
- *Cross-functional:* Processes should span several functions within an organization.

Types of Processes

Business owners, managers, and staff members create processes to outline the specific activities or tasks needed to complete various functions.

This process can involve the use of charts or maps to create a pictorial reference for use by process operators. Owners and managers can break these models down to the lowest task or activity in the company. This allows companies to visualize activities in an end-to-end construct and better reduce duplication and streamline operations. Following are three primary types of processes that can be modeled:

- *Management:* These are business processes that govern the operations of an enterprise or organization. Typical management processes include strategic planning and corporate governance.

- *Operational:* Also known as primary or core processes, these are business processes that form the primary objective of the enterprise and subsequently create the primary value stream. Typical operational processes include Engineering, Purchasing, Manufacturing, Marketing, and Sales. An example of a core process for a manufacturer might be to assemble parts.

- *Supporting:* These are business processes that support the core operational processes of an organization. Typical supporting processes include Finance, Information Technology, and Service Delivery departments such as Project Management and Process Improvement, as well as Organizational Development and Customer/Technical Support. In the manufacturing example, a supporting process might be to recruit production staff.

Process Owner

Process Owners are a critical component of Process Improvement efforts. They are the named individuals responsible for the performance of a process in realizing its objectives. Process Owners are responsible for the comprehensive management of processes within the organization, including approving documents (Process Maps, Procedures, and Work Instructions), determining key performance indicators, as well as monitoring process performance and recommending improvements. A Process Owner is the only person with the authority to make changes to a process and is the contact person for all information related to its performance. In most cases, Process Owners are leaders within an organization but can also be individuals in nonleadership positions. Process Owners are supported by Process Improvement Organizations and ultimately inherit the solutions created by a Process Improvement project team. A Process Owner should

- Be a subject matter expert of the process or the domain in which the process is classified

- Demonstrate process-oriented thinking

- Understand the outcomes and experience needed to achieve customer satisfaction

- Have insight into performance issues or disconnects within the process
- Be a well-respected professional who can positively influence process operators and project team members

Relationships among Policies, Processes, and Procedures

Process Improvement requires that participants understand the distinction between a *Policy*, a *Process*, and a *Procedure* before beginning improvement activities. All three terms address related subject matter but do so with different types of content and at different levels of detail. Each term has a unique purpose that drives the content contained within each document. Although these terms are often used interchangeably, Policies, Processes, and Procedures are in fact distinct items. Consequently, it is important to have a formal definition of each term prior to embarking on improvement initiatives. With this understanding, participants and stakeholders know what work will be performed over the course of a Process Improvement project. Although it is very easy to lump these terms together as if they were a single entity, knowledge of the difference will enable participants to understand the documentation that will be produced, as well as deliver higher-quality artifacts throughout Process Improvement projects.

Policies

Policies are guiding principles that are intended to influence decisions and actions across an organization and govern the implementation of its processes. They include laws, guidelines, strategic goals, and business rules under which an enterprise operates and governs itself (why an organization does something). They contain the formal guidance needed to coordinate and execute activity throughout an organization. When effectively deployed, policies help ensure process designs meet organizational standards. There is not a one-to-one relationship between a Policy and a Process as corporate policies may affect multiple processes. However, processes must reflect the business rules contained in any related Policies. Policies outline a particular principle and its classification, describe who is responsible for its enforcement, and outline why the Policy is required. A simple Policy Document is the most common tool used to describe a Policy (Figure 2-1).

NOTE *A Business Rule is a declarative statement of control or constraint that the business places upon itself or has placed upon it. Policies, on the other hand, are general or informal statements of direction for an enterprise. Although both are considered elements of guidance, policies are usually translated and refined into business rules. A Policy may be the basis for one or more business rule statements, just as a business rule statement may be based on one or more policies. An example of a Policy–Business Rule relationship for a car dealership would be the following:*

Policy Template

Policy Overview	
Policy Title	
Policy #	
Owner	
Effective Date	
Expiry Date	

Policy Description	
Policy Statement	
Scope Statement	
Responsibilities	

Contacts			
Subject	Contact	Telephone	E-mail Address

Definitions	
Term	Definition

Related Documents	
Related Processes	
Other Documents	
Forms and Tools	

Document History				
Date	Version	Change Description	Author	Approval

FIGURE 2-1　Sample Policy

- *Business Policy: We only sell vehicles in legal, roadworthy condition to our customers.*
- *Business Rule: Vehicles must be safety checked upon return from each customer test drive.*

Processes

Processes are related activities or actions that are taken to produce a specific service, product, or desired result. Processes can be formal or informal, large or small, specific to a set of cross-functional departments, or span across an entire organization. Processes are publicly known, documented, supported, and widely used by an organization. They contain resources, steps, inputs, and outputs used to indicate where there is separation of responsibility and control within a series of related and connected activities (what an organization does or should be doing). Unlike the relationship of Policy to Procedure, there is a direct relationship between Processes and Procedures as Procedures describe in detail how each activity within a process is carried out. Processes address who is responsible for performing activities (departments or divisions), what major activities are performed, and when the process is triggered and subsequently halted. Common tools used to display processes include Flowcharts, Cross-Functional Process Maps, and Integrated Definition for Functional Modeling (IDEF) Diagrams.

An example that illustrates this concept is the process of repairing a vehicle at a car dealership. The process is triggered by a customer's request for an appointment and concludes when the vehicle is returned to the customer for use. Figure 2-2 illustrates this in a simple Process Map.

NOTE *A Subprocess is a set of activities that have a logical sequence and that meet a clear purpose but with functionality that is simply part of a larger process. Just like processes, there are goals, inputs, outputs, and owners, but they are modeled as lower levels of detail within larger processes. Subprocesses are important because they play a major role in improving overall processes. In many cases, processes are too large to analyze and improve at one time, and so grouping common process activities into subprocesses can help Process Improvement efforts tremendously. It is recommended to break large processes into separate master processes and child processes, because it reduces the complexity of the process map and allows subprocesses to handle exceptional situations and ancillary activities. Sub-Processes are also useful for hooking the functionality of an existing process into another process.*

Procedures

A *Procedure* is a set of written instructions that define the specific steps necessary to perform the activities in a Process (Figure 2-3). They document the way activities are to be performed to facilitate consistent conformance with organizational requirements and to support quality and consistency. They define how the work is performed and are typically documented in step-by-step fashion, describing in detail each activity within a process. Procedures detail specifically who performs the activity (the role within a department), what steps are performed, when the steps are performed, and how the steps are performed, including any standards

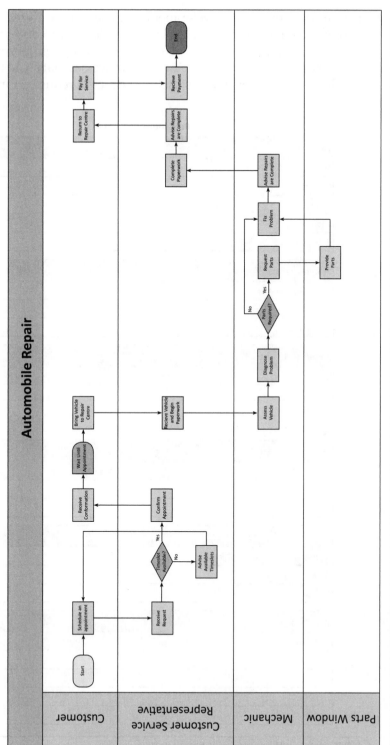

FIGURE 2-2 Simple Process Map

that must be met. The development and use of Procedures is an integral part of successful process-focused organizations. Procedures provide individuals with the information needed to perform their jobs properly and they detail the regularly recurring work processes that are to be conducted or followed. Many businesses use Procedures to reduce errors, assist with

Procedure Template

Procedure Overview	
Procedure Title	
Procedure #	
Owner	
Effective Date	
Expiry Date	

Procedure Description	
Purpose Statement	
Scope Statement	
Responsibilities	

Procedures		
Step	Action	Responsible

Definitions	
Term	Definition

Related Documents	
Internal Documents or Policies	
Other Documents	
Forms and Tools	

Document History				
Date	Version	Change Description	Author	Approval

FIGURE 2-3   Sample Procedure

training employees, or as a point of reference to ensure consistency of work. In doing so, use of Procedures can minimize variation and help ensure quality through consistent implementation of activities within an organization. Procedures can take the form of a Work Instruction, a quick reference guide, or a detailed Procedure Document.

NOTE *Work Instructions are a form of Procedure and are generally recognized as a subset of Procedures. However, they are typically written to describe how to do something for a single role or activity within a process, whereas full-fledged Procedures describe the detail of every activity within an end-to-end cross-functional process. Well-written Procedures can often eliminate the need for documenting Work Instructions. However, most process mapping and modeling tools available today enable and encourage Work Instructions over formal Procedures due to the click-through and component-oriented nature of the applications. Organizations may choose the most optimal and user-friendly route for their users.*

Using Policies, Processes, and Procedures

The easiest way to think about Policies, Processes, and Procedures is in the level of detail contained within each item (Figure 2-4). Processes can be described as being at a high level and operating across an organization's various functions, whereas procedures are at a lower level and contain

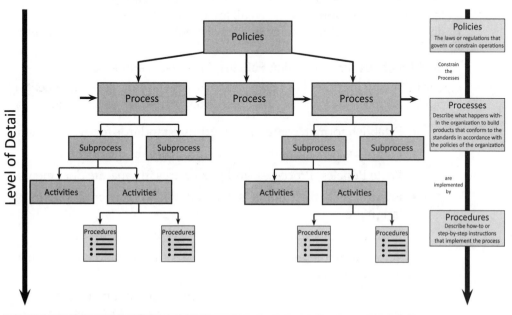

FIGURE 2-4   Process Hierarchy

more detailed information, breaking down the various activities within a process. In simple terms, the primary difference between a process and a procedure is that a process is what you do and a procedure is how you do it. Policies are at the macro level of detail and contain guiding principles or rules intended to influence decisions and actions carried out in Processes and Procedures. All are innately linked and, depending on the organization's needs, may or may not be required in order to execute day-to-day activities or Process Improvement efforts.

Processes are an excellent means of displaying a series of related activities in an easy to understand format; however in many cases they are deemed too high level for employees to use to perform their day-to-day tasks. The most common solution to this problem is to pair process diagrams and procedural detail together, clearly detailing the steps in a process that a procedure refers to. By doing this, employees can see the greater context and implications of the cross-functional process while also having the level of detail required to successfully complete their duties. When proper policies exist, everyone within an organization who is using different processes and procedures will be able to see the connection of what they are doing to why they are required to do so in a certain manner.

The most important element is that there is a cascading effect of improvements made at a higher level on those at a lower level. If a recommendation to replace a given Policy with a better one is made and accepted by business Process Owners, then corresponding changes in the subsequent Processes and Procedures must follow in order to ensure compliance. Policies, Processes, and Procedures are all part of an organization's Process Ecosystem and usually work best when all are managed together. Figure 2-5 outlines the distinguishing characteristics of a policy, process and procedure.

Best Practices for Developing Policies, Processes, and Procedures

The following list describes best practices for developing and implementing Business Processes, Policies, and Procedures for use across an organization:

- Establish format, content, and writing standards for all three items
- Identify who will write, review, and maintain the documents
- Write Policies, Processes, and Procedures in separate documents
- Ensure each document is written with clear, concise, simple language
- Ensure documentation is easy to access, use, and update
- Setup common vocabulary to reduce confusion and ensure consistency of terms
- Use version controls to ensure proper document management
- Ensure subject matter experts are identified and readily available to resolve issues

DISTINGUISHING CHARACTERISTICS

The distinctions commonly drawn between Policies, Process, and Procedures can be subtle, however, there are characteristics that help discern Policy from Process and Procedure. They are:		
Policies	Processess	Procedures
• Policies are driven by environmental factors	• Processes are driven by achievement of a desired outcome	• Procedures are driven by completion of a task
• Policies are adhered to	• Processes are operated	• Procedures are executed
• Policies statements serve to guide process design and operation	• Process activities are completed by different people with the same objectives, department barriers are non-existent	• Procedures steps are completed by different people in different departments with different objectives
• Policies are expressed in broad terms and guide activities	• Processes are less detailed and describe what activities should occur	• Procedures are highly detailed and describe how activities are executed
• Policies focus on satisfying controls	• Processes focus on satisfying customers	• Procedures focus on satisfying rules and standards
• Policies define organizational constraints	• Processes transform inputs into outputs through use of resources	• Procedures define the sequence of steps to execute a task
• Policies are evolutionary	• Processes are dynamic	• Procedures are static
• Policies guide people, actions, and events	• Processes cause things or events to happen	• Procedures cause people to take actions and decisions
• Polices change less frequently	• Processes change gradually	• Procedures are prone to change
• Policies have widespread application	• Processes are moderately spread	• Procedures have narrow application

FIGURE 2-5   Distinguishing Characteristics of a Policy, a Process, and a Procedure

- Develop each artifact with the customer and/or user in mind (internal or external)
- Ensure each document does not include any unnecessary information

Benefits of Developing Policies, Processes, and Procedures

Development of Policies, Processes, and Procedures helps organizations

- Provide visibility into areas of quality, productivity, cost, and schedule
- Improve communication and understanding

- Plan and execute activities in a disciplined fashion
- Capture lessons learned
- Facilitate the execution of organization-wide processes
- Analyze and continuously improve operations
- Reach new maturity levels
- Provide a basis for training and skills assessment

Process Improvement Defined

Process Improvement is an ongoing effort to improve processes, products, and/or services in order to meet new goals and objectives such as increasing profits and performance, reducing costs, or accelerating schedules. These efforts can strive for incremental improvement over time or rapid improvement in a very short period and often follow a specific methodology or approach to encourage and ultimately create successful results. There are several widely used methods of Process Improvement such as *Kaizen (Plan-Do-Check-Act)*, *Lean Six Sigma*, and *Rummler–Brache*. These methods emphasize employee involvement and teamwork by measuring and organizing processes; reducing variation, defects, and cycle times; and eliminating process waste to better service customers and deliver value faster and more effectively.

Process Improvement provides a framework that will facilitate continuous improvement, process design, and performance measurement activities that ultimately drive the fulfillment of important organizational goals. It can be applied to all industries and is relevant to all areas of an organization because processes naturally degrade over time for any number of reasons. In addition, the act of continuously monitoring and improving these processes helps proactively resolve issues in order to avoid operating in a crisis management environment when process degradation occurs.

Process Improvement efforts will almost always result in dramatic, positive returns on investment that affect the organization and its people, practices, and products in a positive way. The primary objective of Process Improvement is to continually improve process productivity. Since process productivity is measured in terms of effectiveness, efficiency, and quality, an effective process is one that produces the right results consistently. Process Improvement enables organizations to

- Increase customer satisfaction
- Reduce unnecessary business costs
- Produce higher-quality products
- Eliminate wasteful activities

- Improve employee morale
- Reduce interdepartmental conflict
- Improve efficiency and effectiveness of operations
- Reduce cycle time and variation
- Break down process silos
- Design more robust solutions

Process Improvement Organization

In today's economy, processes are constantly evolving as organizations seek new ways to reduce costs, improve operations, increase productivity, and eliminate waste. Managing these diverse efforts along with the people, resources, technology, and communication that go along with them is often a challenging endeavor. An effective solution, one that is created to establish a more centralized management structure for Process Improvement efforts, is the Process Improvement Organization. A *Process Improvement Organization* is assigned various responsibilities associated with the centralized and coordinated management of Process Improvement efforts and projects across an enterprise. Responsibilities can range from directly managing Process Improvement Projects, to implementing and modeling business process architectures, to simply providing Process Improvement support to functional managers by way of guidance and education. The function provides organizations with skilled Process Improvement Professionals who provide oversight and leadership services throughout improvement activities, either in a consulting arrangement or as dedicated delivery managers or subject matter experts. They deliver the necessary architectures, procedures, systems, and tools necessary to achieve effective Process Improvement by leveraging various methodologies and standards, allocating resources, establishing consistent performance criteria, and reducing duplication of efforts within an organization.

How a Process Improvement Organization is staffed or structured depends on a variety of organizational factors including targeted goals, traditional strengths, and the cultural imperatives of the organization or enterprise it supports. Process Improvement projects managed by the organization are not necessarily related to one another. However, they must provide some form of value or improvement to the organization and must consider the overarching business process architecture, as well as any customer who might benefit from the endeavor. In general, the Process Improvement Organization provides the framework for the construction and maintenance of an organization's business process landscape and the execution of major Process Improvement projects. It is recommended that all Process Improvement activities be coordinated through this department to ensure the stability and sustainability of organizational processes.

Responsibilities of a Process Improvement Organization

The Process Improvement Organization has many responsibilities including capturing and documenting processes, managing and facilitating workshops, gathering performance data for management review, capturing improvement requests, and keeping an up-to-date repository of all processes and related documentation. Other responsibilities include

- Identifying and implementing appropriate Process Improvement methodologies, best practices, and standards tailored to a corporation's needs
- Developing and managing project checklists, guidelines, templates, and other best practices
- Managing Process Improvement activities and projects throughout the entire Process Improvement life cycle
- Managing and facilitating Process Improvement workshops
- Designing and building business processes and overseeing the enterprise process ecosystem
- Administering and maintaining organizational process models and repositories
- Coaching, mentoring, and training business partners and management staff
- Ensuring alignment of process improvements with strategic business goals and priorities
- Monitoring performance of processes for opportunities of improvement and areas of risk

Benefits of a Process Improvement Organization

There are many benefits to establishing an effective Process Improvement Organization including

- *Increased efficiency when delivering improvements:* Methods used by Process Improvement Organizations provide a roadmap that is easily followed and help guide stakeholders to project completion.
- *Heightened customer satisfaction:* Formal Process Improvement efforts that are completed to satisfaction, that are kept to a defined schedule, and that deliver valuable results increase customer satisfaction.
- *Service predictability:* Improvement efforts managed out of a central organization provide predictable service and bring a sense of comfort to customers.

- *Increased process focus:* Positive improvement results that are carried out by project teams often inspire functional employees to look for ways to perform activities more efficiently.

- *Increased agility:* Strategic planning combined with disciplined Process Improvement enables flexibility through structured processes that are designed to continually adapt to changing organizational conditions.

- *Broader context:* Improvement conducted through a centralized governing body allows all changes to be analyzed for downstream effects or potential business conflicts.

- *Mature service delivery:* A structured delivery approach ensures proper standards are in place and provides more mature management of Process Improvement projects.

Process Improvement and Operations Management

The coexistence of Process Improvement and *Operations Management* can be a complex relationship at times, particularly when improvement efforts or projects require the implementation of large-scale changes or alterations to day-to-day operations. A smooth and amicable relationship between these two functions is critical. Before embarking on any major improvement project, all interested parties should become well versed in the differences between Process Improvement and Operations Management as well the responsibilities of each function. An *Operational* or *Functional Department* is any group of individuals that carries out a set of repetitive tasks or processes to satisfy mission-critical requirements or to produce products or services. These are the day-to-day activities that are required in order to sustain the business and are permanent in nature. Examples of operational functions include Order Processing, Legal, Manufacturing, Accounting, Finance, and Sales.

Process Improvement is the ongoing effort to improve an organization's processes and activities. When operational processes or tasks require improvements in order to achieve better results or meet particular strategic objectives, unique and temporary business initiatives or projects are formed. These efforts require the expertise and knowledge of a dedicated Process Improvement professional to ensure alignment and disciplined and structured delivery.

A *Process Improvement Project* is an individual or collaborative initiative designed to determine the cause of a business or an operational issue, define and analyze current performance of activities related to the issue, implement improvements that rectify the issue, and ensure that appropriate operational transition and monitoring occurs. Process Improvement projects are temporary in nature, have a definitive start and end point, and conclude when projects achieve their improvement objectives or are terminated. Improvement efforts are undertaken to produce specific outcomes such as a reduction in order cycle times, a reduction in unnecessary shipping costs, production of higher-quality products, or elimination of redundant activities.

Once the improvement initiative delivers its output, it is concluded and deployed back into business operations. Although Process Improvement involves continually examining processes for efficiencies and making improvements to enhance the business, an operational unit often initiates these activities when issues or concerns relating to operational performance are identified. The efforts to improve or rectify these issues are then led and managed by a Process Improvement Manager. There is generally significant interaction between an organization's operations department and Process Improvement teams. Process Improvement efforts intersect with operations at many points in time including when

- Performance issues arise in the execution of a process and Root Cause Analysis (RCA) is required
- Activities are not shared or visible, and mapping activities are required
- Processes are not consistent, and operational activities require standardization
- Changes or improvements to an operational process are required
- Direct oversight is required to manage improvements throughout each phase of the Process Improvement life cycle
- Lessons learned or other facilitation is required to discover, change, or discuss better ways of operating

Differences between Improvement Projects and Operations

Key differences between Process Improvement Projects and Operations are

- Improvement Projects are temporary and unique, while Operations are ongoing activities with repetitive output.
- Process Improvement requires project management or facilitation of activities, whereas Operations require ongoing functional management.
- Projects are executed to start a new business objective and terminated when the objective is achieved, while Operational work is performed to keep an organization functioning and to be able to able to sustain the business.
- Projects have a definitive beginning and end, while Operations are ongoing.

- Projects conclude when a unique improvement or result has been attained, while Operations produce the same product service or result on an ongoing basis.

- A Project concludes when its specific improvement goals or results have been attained, whereas Operations takes over any newly designed activities and continues their execution.

Similarities between Process Improvement Projects and Operations

Improvement efforts and Operational activities are

- Performed by people
- Constrained by limited resources
- Planned, executed, and controlled
- Executed to achieve particular objectives

Role of a Process Improvement Manager

A *Process Improvement Manager* is responsible for the development and evolution of an organization's capabilities. This is done by teaching Process Improvement skills and managing any Process Improvement projects or related endeavors. They demonstrate best practices associated with the Process Improvement discipline and inspect business processes for areas that could be improved while creating a positive learning environment for the various stakeholders involved with an organization's processes. They support the delivery of new processes and solutions that are more robust, customer centric, and lean in design. The role of a Process Improvement Manager is distinct from a functional manager or operations manager. The operations manager is typically focused on providing management oversight to a facet of the core business on a permanent basis, whereas a Process Improvement manager is responsible for facilitating the improvement of operational processes throughout the temporary nature of a formal project. Depending on the organizational structure, a Process Improvement Manager may report to a functional manager or into a centralized Process Improvement Organization. The recommended approach is to keep separation of duty in place and have all Process Improvement professionals report into a centralized organization. However, in many cases, organizational cultures, finances, or resources do not permit this. If that is the case, it is still ideal to ensure that the Process Improvement Manager's role is a distinctive role even when inserted into a functional or core operations division.

Process Improvement Managers facilitate change to the various process areas such as cycle time, waste reduction, quality improvement, increased awareness of risk, mitigation of risk, effective communication, and increased customer and business satisfaction. They must be highly skilled in the identification and resolution of strategic challenges and tactical execution and be able to demonstrate these skills in order to mentor other Process Improvement professionals or business stakeholders. The role is typically highly visible in the organization as it is engaged in meaningful corporate change activities at the strategic level. Process Improvement Managers act as advisors and coaches to business owners. In this capacity, they typically do not take on direct responsibility for business team members because it is necessary to ensure that the purity of their purpose is maintained and the integrity of proposed solutions is upheld. Process Improvement Managers should always be collegial and act as a helping hand rather than as a control point. Although operations managers are responsible for managing the day-to-day activities of a process, Process Improvement Managers focus on improving those activities in order to sustain business competitiveness and performance. Process Improvement Managers ensure that any improvement effort is managed in a disciplined and structured fashion and that all efforts align to an organization's strategic objectives.

Process Orientation and Organizational Structure

Organizations require a purposeful set of methods for achieving goals and ensuring processes align with strategic objectives that meet and exceed customer expectations. In order to do this successfully, organizations must determine the structure in which they wish to operate. One type of organizational management structure is the *Functional Management* structure. Functional Management is the most common type of organizational structure and is defined as the configuration of an organization into departments or silos on the basis of the type of work to be performed (e.g., Order Processing, Sales, Demand Management, Logistics, and Manufacturing). In such a model, the Sales Team takes the role of promoting the finished product in the marketplace, Demand Management focuses on predicting future sales, Order Processing focuses on handling orders, Manufacturing focuses on rolling out the final product, and Logistics focuses on shipping orders. Figure 2-6 illustrates the basic structure of a sample functional organization.

While there are inherent advantages to operating in a functional capacity, it is important to be aware that functional organizations are prone to have any number of the following drawbacks:

- The focus is on individual department improvement rather than cross-functional and integrated improvement that focuses on the customer.

- The decision-making process can be bureaucratic and far from expedient when trying to solve customer issues that cross boundaries of multiple functions.

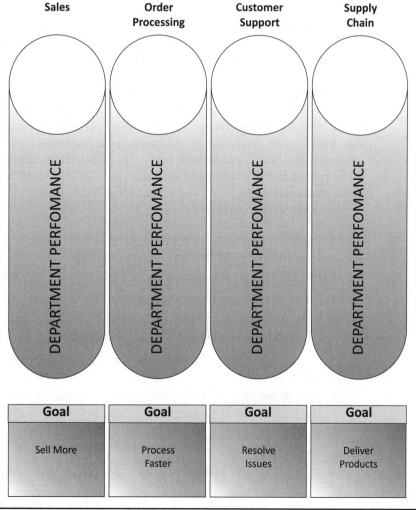

FIGURE 2-6 Sample Functional Organization Design

- The flow of communication and synchronization between functional departments is complicated and typically lengthy.
- The speed of resolving problems can be slow and inefficient.
- Grouping based on functions results in a lack of a broader view of company objectives for employees.
- Successful performance is defined as performing within and winning for the function, sometimes at the expense of the overall organization.
- Things often fall through the cracks or into white space, causing fragmented or duplicate efforts.

NOTE *Rummler–Brache created the term White Space. It refers to the area between the different functions or departments of an organization. In White Space, rules are often vague, authority fuzzy, and process ownership unclear, resulting in misunderstandings and delays.*

Another organizational management structure is the *Process Management Structure*. The Process Management Structure is defined as a management system that facilitates Process Improvement and process design activity in a way that focuses on customer requirements. It promotes common language and goals across all departments and emphasizes cross-functional improvement activities. This management structure is the basis for all Process Improvement efforts.

Process-focused organizations

- Place considerable focus on linking processes between individual functions for end-to-end process coverage
- Ensure everyone is able to more efficiently utilize resources
- Provide a common language across departments
- Provide lessons learned supported by reliable data that can be applied to other processes and improvement activities
- Ensure all employees understand process steps and how they add value
- Ensure employees understand how processes are performing and the value behind capturing this data
- Ensure employees help manage each other instead of escalating conflicts
- Hand-offs between employees are smooth and without artificial boundary
- Ensure processes are objectively and frequently measured and reviewed for performance fit
- Ensure customers' requirements are known by everyone in the organization

The purpose of implementing a Process Management structure is to integrate and harmonize functional and process management systems so that an organization's primary focus is on aligning departments to common goals, which focus on the customer experience. This practice fundamentally shifts the mindset of an organization and helps foster a culture that is process driven and focused on meeting customer requirements and creating predictable results. This philosophy incorporates a "what is good for the customer can also be good for the company" approach.

Figure 2-7 illustrates an example of a process-focused organization. The illustration shows the linkage of process deliverables to customers,

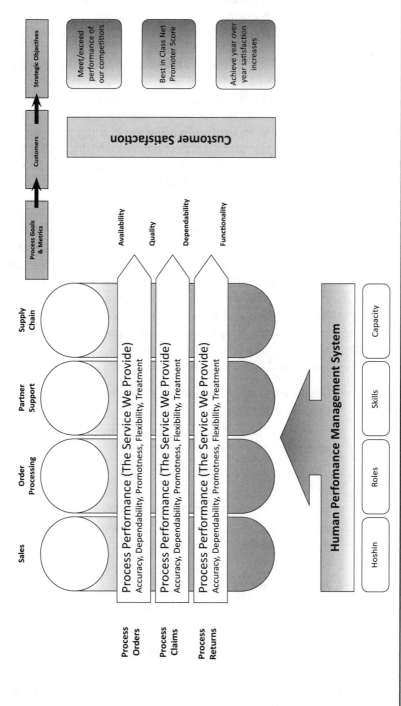

Figure 2-7 Sample Process-Focused Organizational Design

41

how they are measured, the connection of core processes with strategic objectives, and the alignment of strategic objectives with customers. In order for a Process Improvement Organization to be successful, the traditional functional thought process must change to focus more on process performance that delivers value to the customer.

Implementing a Process Improvement framework, eliminating departmental barriers, and becoming process focused takes time as results are delivered incrementally over time. As the shift occurs from well-defended functional silos to cross-functional processes, the organization matures and is able to respond more flexibly, efficiently, and effectively to customer demands. Departments become less divided as the process and its outputs become more visible and dominant. A lean process takes less time, has steps that only add value for the customer, and has fewer hand-offs and errors.

Common changes that a team should plan for when implementing a Process Improvement framework include shifts in employee roles, shifts in department policies, changes in customer requirements, shifts in employee empowerment and engagement levels, some anxiety about an employee's role in the new operation, changes in performance targets, and replacement of existing tools and technologies. Change management is always an important consideration when working with processes, and it is even more essential when adopting a process-driven approach to an operation. Other kinds of changes that occur when an organization streamlines into a process-focused approach are

- Organizational structure changes from functional orientation to process orientation.
- High walls between departmental functions change to cooperative partnerships.
- Operational activities change from simple tasks to multidimensional work.
- Employee roles change from being controlled to being empowered.
- Processes change from being somewhat flexible and inefficient to being highly flexible and efficient.
- Focus on delivering value to a function shifts to delivering value for the customer.
- Quality changes from being add-on to being built-in.
- Managers change from supervisors to coaches.
- Processes change from being uneven to being balanced and capable.
- Focus of performance measures shifts from activity to results.
- Advancement criteria change from being performance based to being ability based.

Leadership and Process Improvement

All organizations realize that leadership is a vital component in achieving their missions. Effective leadership can bring many advantages such as

- Increased employee productivity
- Improved cost management
- Improved resource allocation
- Better change management

However, making sure that Process Improvement efforts take advantage of the many benefits of good leadership requires active communication and involvement. Leadership commitment and support are critical for successful Process Improvement endeavors (Figure 2-8). The following list outlines several support strategies that leaders are encouraged to use in order to properly support Process Improvement efforts:

- *Change organizational norms:* Leaders should strive to change the way people in the organization think. Having people modify their perspectives and adapt new organizational norms is a fundamental activity in shifting to process orientation. All process and system changes require some sort of change effort, and any meaningful improvements cannot occur until organizational norms are changed and individuals embrace them and alter their behaviors accordingly. Leaders should use the persuasive power that comes with their role to create a culture that embraces change instead of fighting it.

- *Set clear objectives:* Leaders should articulate a vision for what process excellence looks like. They should communicate this vision to everyone in the organization and continuously deliver a consistent message regarding priorities to Process Improvement stakeholders. Another critical component is assisting Process Improvement teams in forming a manageable scope for each Process Improvement project. Whether this is through direct involvement or through committee, leaders should help define the attributes needed for success and

Leadership Strategies for Process Improvement	
1. Change Organizational Norms	6. Participate in Activities
2. Set Clear Objectives	7. Implement Feedback System
3. Select Improvement Priorites	8. Monitor Performance
4. Provide Support	9. Remove Roadblocks
5. Dedicate Resources	10. Celebrate Achievements

FIGURE 2-8 Leadership Strategies for Process Improvement

empower employees to develop efficient and effective approaches to accomplish Process Improvement projects.

- *Select improvement priorities:* Leaders should set appropriate targets for the organization and identify the processes they feel are critical for improvement. They should solicit input from process operators and customers in order to identify processes that are in need of attention and determine those that have the greatest potential for improvement. These processes should bring about a sense of urgency in team members and be included in any annual strategic plans for the organization.

- *Provide support:* Leaders should communicate their support for Process Improvement efforts both verbally and through actions on a continuous basis. This can consist of written communication through e-mail to staff articulating the vision, or expressing expectations for these efforts in various leadership meetings such as town halls and team meetings. Following any improvements, leaders should express gratitude for the efforts, demonstrate conformity to the new way of operating, and actively participate in project retrospectives in order to recognize progress and reinforce the importance of continuous improvement.

- *Dedicate resources:* Leaders should dedicate resources and money to support improvement efforts and ensure that the new processes are implemented as planned and that Process Improvement gains are realized. They should work with the Process Improvement Organization and any process owners to ensure that the allocation of staff and funds is suitable for the scale of effort being undertaken.

- *Participate in activities:* Leaders should attend Process Improvement meetings and invite participants to report on progress and share any meaningful successes. They should encourage Process Improvement team members to escalate issues that need resolution and address any concerns promptly. The involvement of passionate and committed senior managers is critical to ensuring the long-term success of Process Improvement efforts.

- *Implement feedback systems:* Leaders should implement feedback systems in order to establish close communication with customers, employees, and suppliers. They should develop methods for obtaining and evaluating process owner, process operator, customer, and supplier input in order to determine any cultural, policy, or procedural obstacles to success.

- *Monitor performance:* Leaders should identify metrics and information that is needed to understand how the organization and its processes are performing. Process Improvement teams should concisely report metrics and information on key aspects of processes, and leaders should review performance metrics on a regular basis. Leaders should also discuss process owner and department management performance in supporting specific Process Improvement efforts during performance reviews. In addition, they should use this information for compensation and promotion decisions, where appropriate.

- *Remove roadblocks:* Leaders should make themselves available for issue resolution and create time during meetings with managers and staff to discuss performance of processes overall and any impediments related to improvement efforts. Leaders should also make themselves visible by walking around the office and engaging employees at their work stations to ask specific questions about how the processes and subsequent improvement projects are working, what support might be required, and what challenges are being experienced. Where roadblocks cannot be removed, leaders should work with managers to build alternate strategies.

- *Celebrate achievements:* Leaders should recognize Process Improvement accomplishments whenever possible and communicate this to staff at meetings and via newsletters, e-mails, and through internal or external corporate websites. Other options include awards and parties to acknowledge individual and team achievements. The more leadership acknowledges Process Improvement efforts, the more engaged and motivated team members become.

Summary

In this chapter, we introduced several key terms and concepts associated with Process Improvement. This chapter outlined the relationships among Policies, Processes, and Procedures; described the Process Improvement Organization; and introduced the concept of a Process-driven enterprise. We also learned that

- Processes are related activities or actions taken to produce a specific service, product, or desired result.

- Three primary types of processes can be modeled: Management Processes, Operational Processes, and Supporting Processes.

- Process Owners are a critical component to Process Improvement efforts. They are the named individuals responsible for the performance of a process in realizing its objectives.

- Policies are guiding principles intended to influence decisions and actions across an organization and to govern the implementation of its processes.

- A Procedure is a set of written instructions that define the specific steps necessary to perform activities in a Process.

- Several best practices are recommended when developing and implementing Business Processes, Policies, and Procedures for use across an organization. These include establishing standards for all three items; identifying who will write, review, and maintain each item; and writing each in separate documents.

- Process Improvement is an ongoing effort to improve processes, products, and/or services in order to meet new goals and objectives, such as increasing profits and performance, reducing costs, and accelerating schedules.

- A Process Improvement Organization is assigned various responsibilities that are associated with the centralized and coordinated management of Process Improvement efforts and projects across an enterprise.

- A Process Improvement Manager is responsible for developing the capability of an organization by teaching Process Improvement skills and managing any Process Improvement initiatives or endeavors.

- Functional Management is the most common type of organizational structure and is defined as the configuration of an organization into departments or silos on the basis of the type of work to be performed.

- Common changes that a team should plan for when implementing a Process-Oriented approach include shifts in employee roles, shifts in department policies, shifts in employee empowerment and engagement levels, and changes to performance targets.

- Leadership commitment and support are critical for successful Process Improvement endeavors. Several support strategies can be employed to ensure successful Process Improvement implementation.

Chapter Preview

Now that the basic fundamentals and key terms of Process Improvement have been discussed, the next step is to continue learning the key knowledge areas of Process Improvement. Chapter 3 covers Process Maturity in detail and outlines the key components that make up a mature organization. A roadmap for evaluating or assessing process maturity within an organization is also discussed.

CHAPTER 3
Process Maturity

It is important for corporations to monitor and assess organizational process maturity. This helps a corporation identify potential areas of improvement and highlights business-critical functions that, if not at an appropriate maturity level, may put a business at risk. However, many organizations become confused with the plethora of standards by which to assess the maturity of their business processes. As a result, they are unable to properly assess the risk that immature processes pose to enterprise initiatives. They are also unable to identify the causes of weaknesses in their operating environments that, if addressed, could reduce cost and increase operating efficiencies. This chapter outlines the components that make up the Process Maturity Model (PMM) and describes the various levels of process maturity at which an organization may be operating. It provides a concrete model for evaluating and assessing the maturity of processes and related improvement efforts within a department or an organization and identifies several common signs of process immaturity. This chapter is organized around the following topics:

- *Process Maturity Model defined:* What is the Process Maturity Model and what is it used for?

- *Process Maturity levels:* What are the various process maturity levels? What are the differences between each level of process maturity?

- *Assessing Process Maturity:* When is an organization ready to assess its maturity? How is process maturity typically assessed?

Process Maturity Model Defined

The *Process Maturity Model* (*PMM*) is a set of structured levels that describe how well the behaviors, practices, and processes of an organization can reliably and sustainably produce required outcomes. The model is based on the process maturity framework first described by Watts Humphrey in his 1989 book *Managing the Software Process* and later enhanced and popularized by Carnegie Mellon University. It describes an evolutionary improvement path that guides organizations as they move from immature, inconsistent business activities to mature, disciplined processes.

The model orders these stages so that improvements at each level provide a foundation on which to build improvements undertaken at the next level. The basic concept underlying process maturity is that mature organizations do things systematically, while immature organizations achieve their outcomes as a result of heroic efforts put forth by individuals using tactics that they create spontaneously. Ultimately, Process Maturity is an indication of how close an organization's processes are to being complete and capable of continual improvement through qualitative measures and feedback. Thus, for a process to be mature, it has to be complete, useful, known by all participants and stakeholders, automated where applicable, reliable in information, and continuously improved. The Process Maturity Model also assists organizations in

- *Deploying Process Improvement frameworks:* The Process Maturity Model is used to guide improvement departments, programs, and initiatives.

- *Evaluating organizational capability:* The Process Maturity Model provides a standard against which to evaluate the capability of processes in meeting service levels, quality, cost, and functionality commitments.

- *Conducting organizational benchmarking:* The Process Maturity Model allows organizations to evaluate their standing relative to the maturity of business processes within organizations across their industry segment.

- *Identifying appropriate areas for improvement:* The Process Maturity Model is used to determine where weaknesses may exist in departments and/or processes and enables organizations to adequately assign recourses to improve said areas.

- *Selecting high-priority improvement actions:* The Process Maturity Model allows organizations to prioritize organizational process enhancements based on a structured and proven road map.

Process Maturity Levels

The Process Maturity Model contains five levels that are used to describe the state of process maturity within an organization (Figure 3-1). Within the model, maturity is measured on an ordinal scale, and each level is used for benchmarking and evaluation and describes the key stages needed in order to achieve a fully effective Process-Oriented environment. The five levels of the Process Maturity Model are described as follows:

- *Level 1—Informal:* Processes at this level are usually undocumented and in a state of dynamic change, tending to be driven in an ad hoc uncontrolled or reactive manner by operators or business events.

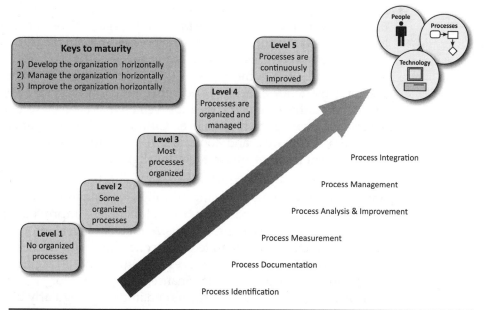

Keys to maturity

1) Develop the organization horizontally
2) Manage the organization horizontally
3) Improve the organization horizontally

Level 5
Processes are continuously improved

Level 4
Processes are organized and managed

Level 3
Most processes organized

Level 2
Some organized processes

Level 1
No organized processes

People
Processes
Technology

Process Integration

Process Management

Process Analysis & Improvement

Process Measurement

Process Documentation

Process Identification

FIGURE 3-1   Process Maturity Levels

This provides a chaotic or an unstable environment for processes where emphasis is often on doing whatever it takes to get things done. At this level, Process Improvement practices are performed inconsistently in pockets across the organization, and processes are often successful because of heroic efforts carried out by process operators.

- *Level 2—Documented:* At this level, some processes within an organization are documented, possibly with repeatable and consistent results. Process discipline is unlikely to be rigorous. However, where it exists, it may help to ensure that existing processes are maintained during times of stress. At this stage, best practices are starting to emerge and some basic measurements are drawn. Implementing a corporate culture that supports the methods, practices, and procedures of Process Improvement has also started at this level.

- *Level 3—Integrated:* At this level, there are sets of defined and documented standard processes established and subject to some degree of improvement over time. These standard processes are firmly in place (i.e., they are the current processes of the organization) and are used to establish consistency of process performance across the enterprise. Process Management and Improvement practices are beginning to be integrated across the organization and include those that enable the successful management of significant, high-risk, complex projects.

- *Level 4—Managed:* At this level, process metrics are established and used to effectively control As-Is processes. Organizations are able to identify ways to adjust and adapt processes without measurable losses of quality or deviations from specifications. Processes are managed with some cross-functional integration and show greater consistency of actions and better communication between functions. At this level, Process Improvement has been elevated to a strategic management practice. Cultural and organizational behaviors, structures, and tools are in place to ensure Process Improvement projects are strategically aligned.

- *Level 5—Optimized:* At this level, the focus is on continually improving process performance through both incremental and innovative business and technological changes/improvements using established metrics and controls. At maturity level 5, processes are cross functionally integrated and concerned with addressing statistical common causes of process variation and changing the process to improve process performance. The goals set for processes are being analyzed for achievements and improved regularly using Process Improvement techniques such as Six Sigma and Kaizan. The Process Ecosystem is being improved and efforts are underway to error-proof processes using techniques such as Poka-Yoke.

NOTE *Poka-Yoke is a Japanese term that means mistake proofing. It is considered as any mechanism inserted into a process that helps a process operator avoid mistakes when executing a process activity or series of activities. Its purpose is to eliminate defects by preventing, correcting, or drawing attention to human errors. The concept was formalized as part of the Toyota Production System management and improvement philosophy.*

Several characteristics of each process maturity level are outlined below.

- Maturity Level 1
  - Performing only core duties
  - Working extended hours to complete day-to-day activities
- Maturity Level 2
  - Defining a Standard Process
  - Performing the Defined Process
- Maturity Level 3
  - Planning performance
  - Ensuring disciplined performance
  - Verifying performance
  - Tracking performance

- o Training employees
- o Utilizing Process Improvement and Project Management
- Maturity Level 4
  - o Establishing measurable quality goals
  - o Managing performance objectively
  - o Preventing defects
  - o Performing change management
- Maturity Level 5
  - o Improving organizational capability
  - o Improving process effectiveness
  - o Preventing defects

Assessing Process Maturity

Process Maturity Assessments are a diagnostic tool used to appraise and characterize an organization's processes. This includes all business activities and supporting tasks, tools, methods, practices, and standards involved in the development and improvement of a process throughout the Process Improvement life cycle. It has become widely accepted that the quality of an organization's products and services is largely dictated by the quality of the processes used to develop and maintain them. As a result, formal Process Maturity Assessments are used to identify the strengths and weaknesses of an organization's processes, highlight organizational risks, and determine capability and maturity level ratings. These assessments assist with

- Measuring the ability of the organization's collective staff to repeatedly deliver outcomes that meet specifications on time and within budget or other specifications
- Identifying gaps in current operational or Process Improvement capabilities
- Providing a foundation for improvement and guidance for advancement through prioritized, structured, and sequential improvements
- Providing an indicator of how effective the client's organization is in meeting its goals in managing projects and meeting business objectives

NOTE *The scope of a Process Maturity appraisal can be any portion of an organization, such as the entire company, a specific business unit, a specific geographic region, or a specific project.*

An organization is ready to conduct a process maturity assessment when it becomes dissatisfied with any aspect of its current performance. Although the desire to improve is a natural first step, an organization must also be willing to demonstrate support for the assessment at the leadership level, and there must be a strong organizational commitment to any subsequent improvements or transformation efforts. Organizational resources and process operators must also be committed to not only conducting the assessment but also to supporting the improvement plans that follow. Assessments should not be undertaken until the affected organization has been informed and prepared for the assessment, necessary communications have been sent, and the need to measure improvement or to reassess at periodic intervals is supported by both process owners and organizational leaders.

Common signs that an organization is ready for formal assessment include

- Roles and responsibilities are not clearly defined within the organization.
- Employees do not understand the order in which work should be accomplished.
- Documentation about how to perform activities is not readily available.
- A process-oriented culture does not exist.
- Processes are not executed properly, resulting in errors, rework, mistakes, or scrap.
- Management assigns resources to fix problems, but the problems are never fully solved.
- Employees are often frustrated while working, and morale is low.
- Customers (either internal or external) are continually unhappy with performance or results.
- Processes are disjointed across several departments, and finger-pointing and blaming exist when errors occur or performance worsens.
- Processes are not controlled or measured.
- A Customer-driven culture does not exist.
- Processes and activities contain too many reviews and sign-offs.
- Complexity, exceptions, and special cases are common.
- Established procedures are circumvented to expedite work.
- No one owns or manages processes.
- Teams spend a great deal of time rectifying issues.
- There are inconsistent procedures for identical tasks within the organization.

Overall, Process Maturity Assessments are an effective strategy for making Process Improvement a priority within an organization. They provide organizations with the information necessary to understand their position as a process-oriented enterprise, help build a clear picture of performance capability, and provide a comparison to competitors. They also provide answers to such questions as

- What are our company's strengths and weaknesses?

- How can our Process Improvement practices enable us to achieve even greater profit/shareholder value while meeting our strategic goals and objectives?

- What areas do we need to improve so that we can immediately increase profitability?

- Do we need to modify our existing processes or practices, add new tools and technologies, or provide additional training for our staff?

- What is the overall morale or engagement of our organization? How can we better motivate, coach, and lead employees?

NOTE *Although similar, Assessments and Evaluations are two distinct methods for determining Process Maturity. The primary difference between a maturity assessment and a maturity evaluation is that an assessment is an appraisal that an organization does to and for its own operations, whereas an evaluation is an appraisal conducted by an external group that comes in and looks at an organization's process capability to help the organization make a decision regarding its future. After an assessment, the assessed organization's leadership owns the assessment findings and results and typically uses the results to formulate an action plan for improvement. After an evaluation, the sponsoring organization's leadership owns the evaluation results, but uses the results and recommendations of the external firm to make decisions regarding process structure, performance, risk, and improvement. In many cases, the external firm also drafts and proposes an action plan for the receiving organization. Assessment results are usually communicated across the organization, whereas the results of an Evaluation may or may not be shared with the evaluated organization.*

Approach

Effective maturity assessments gather multiple, overlapping forms of evidence to evaluate the performance of an organization's processes and activities, including

- Artifacts that are produced by performing a process

- Artifacts that support operating or performing a process

- Interviews with individuals or departments that operate processes

- Interviews with individuals who own or oversee a process and its performance
- Interviews with individuals who support a process
- Data used to exemplify the current state of organizational culture and/or the attitudes and behaviors found within it
- Data used to describe the performance of a process, its outcomes, and business results

Ultimately, Process Maturity Assessments are multidimensional and include the administration of surveys, document reviews, and a series of individual and small group interviews where results and subsequent recommendations are documented and summarized into reports. The following sections outline the typical assessment process.

Assessment Team

An assessment team is usually comprised of two assessors: one senior professional who serves as team lead and one mid-level professional who serves as an assessment team member. The lead assessor can be internal to the organization or an outside consultant as long as the individual is experienced in conducting assessments and is thoroughly grounded in the Process Maturity Model. An external assessor is likely to have the advantage of more assessment experience and the kind of perspective that arises from having conducted assessments across a broad spectrum of organizations. Assessment team members should be completely independent of the organization, processes, or projects under review in order to minimize the chance of positive bias, which often occurs when internal assessors are used. Experience performing various assessments has proven that it is critical that assessments be led by qualified and well-trained individuals in order to achieve accurate results in a reasonable amount of time and achieve a successful organizational intervention. By having well-trained and qualified assessors, different assessors should get similar results in organizations of similar maturity.

NOTE *To ensure that Process Maturity Assessments have a successful impact, many factors need to be carefully considered. Foremost, the person who leads the assessment team must be well qualified and trained in assessment techniques and Process Improvement frameworks, methods, and concepts. In addition, members of the assessment team must also meet certain criteria such as having industry or organization knowledge. Furthermore, the selection of assessment participants must adequately cover the scope of the assessed organization.*

Assessment Planning

Assessment planning begins with identification of the goals to be addressed by the assessment. These goals should be clearly communicated to the organization at the outset of the maturity assessment. Preparation also involves selecting the various processes and teams to be analyzed, planning and scheduling assessment activities, customizing survey questions, and scheduling and preparing for the kickoff meeting, which is held with all assessment participants. An important objective of this meeting is to prepare the assessment participants for what will occur throughout the assessment period and answer any questions they may have.

Survey Administration

Survey administration involves the completion of assessment surveys by the process owner, key process operators, and selected subject matter experts for each department and process involved in the assessment. The purpose of the survey is to gather written information that is used to determine whether a particular organization is or is not performing at a desired level. An assessment team is then able to summarize the results for a quick overview of the organization's strengths and weaknesses. It is recommended that survey distribution be limited to those involved, impacted, or related to the processes in question, as the process-oriented language used in the survey may not be understood by all individuals across the organization.

Interview Sessions

Individual interviews are conducted with leaders across the organization, Process Improvement Organization members, process owners and operators, and other key stakeholders to validate survey responses. Group input can be obtained during the interview process in one-on-one meetings or focus groups to gain additional insight into the organization's process performance and Process Improvement practices. Interviews serve to validate survey responses by determining the participant's understanding of the processes being evaluated and the accuracy of responses provided by the various survey participants.

Artifact Reviews

Process artifact inspections and Process Improvement project deliverable reviews are completed to further validate survey responses and interview results. Document reviews help establish context and reveal areas that the assessment team needs to probe. A review of the documents also helps the assessment team evaluate and understand the processes that an organization uses in order to rate their strengths and weaknesses relative to the Process Maturity Model. Artifact reviews should include process documents such as policies, process maps and procedures, project-level documents such as project charters and schedules, as well as performance documents such as dashboards and reports.

Maturity Rating

The assessment team must assign an organizational maturity rating based on the results from the assessment surveys, interviews, and focus groups, as well as artifact inspections. Ratings by the assessment team depend on the quality of the data available to them, their ability to reliably judge the implementation and institutionalization of the process improvement practices in the organization, and their ability to correlate these practices with the Process Maturity Model.

Validation and Prioritization

Based on the assessment findings and overall maturity rating, the team develops a set of recommendations that can be presented to and used by the organization in question. It is important for assessment participants to validate and prioritize the team's findings prior to finalization of the assessment deliverables. The employees who participated in the assessment should set time aside to validate findings and prioritize recommendations. This review technique helps the assessment team focus the final recommendations on those offering the greatest improvement impact for the organization.

Assessment Report Preparation and Delivery

The assessment team analyzes their findings based on all feedback and prepares the final assessment report. The improvement recommendations contained within the report should be presented informally to management prior to communication with any other group. This allows management the opportunity to not only understand the recommendations but to commit to implementing them in a timely manner. The recommendations are then presented formally in a findings presentation to the project team, key stakeholders, and all assessment participants. Assessment deliverables should be published and distributed to stakeholders immediately after the findings presentation. Deliverables typically include the findings and recommendations report, the survey results report, and an executive presentation.

Timeline

The amount of time required to conduct a formal Process Maturity assessment depends on the organization's size, complexity, and culture. However, most assessments are typically completed within four weeks. A common structure and timeline for conducting assessments is outlined below.

Planning—Duration, 1 Week. During the first week, the assessment team is formed and the organization in question is educated. Time is also spent customizing the assessment approach for organization-specific requirements and preparing the organization for the assessment launch.

Activities include preparing an assessment plan, scheduling workshops and interviews, and completing survey and interview questions. A kickoff meeting is conducted at the conclusion of week 1 to formally launch the maturity assessment process. Other key activities that are part of the planning phase include

- Verifying the organization's or sponsor's business goals for the assessment

- Determining organizational scope (the areas of the organization that will be assessed and any selected projects and assessment participants)

- Building a schedule for assessment activities and identifying the resources that will be used to perform these activities

- Mapping the assessment outputs and any anticipated follow-on activities

- Tailoring the assessment method for the organization in question

- Outlining any risks or constraints associated with executing the assessment

- Obtaining sponsor endorsement of the plan and authorization for the assessment to be conducted

Interviews and Surveys—Duration, 2 Weeks. During weeks 2 and 3, the assessment team completes the surveys and interviews with various members of the organization. The assessment team begins to review survey responses and prepares and conducts interviews with various process owners and operators. The assessment team also conducts interviews with management and small groups of practitioners. Facilitated focus groups and lessons-learned sessions are conducted; the assessment team also reviews various project deliverables to corroborate findings and gather as much information about the organization as possible. Confidentiality of all participants and data sources is critical to maintaining stability and credibility.

Recommendations and Improvements—Duration, 1 Week. The assessment team collaborates and discusses the various findings from survey responses, interviews, and artifact reviews. The team then drafts findings in terms of strengths and recommendations to advance process maturity. A maturity rating is also assigned at this stage, and results are often validated and prioritized by those who participated in the assessment. Last, the assessment team prepares and delivers an assessment report that outlines the current state of organizational practices, proposes recommendations for improvement, and provides the organization's overall maturity ranking.

Results

A Process Maturity Assessment could reveal significant deficiencies or areas of improvement for an organization. As a result, a disciplined plan should be in place to build on the momentum of the assessment and to follow up on its findings and recommendations. This is achieved by putting in place a rigorous plan for implementing necessary changes. As time passes, the assessment's momentum and power to motivate change diminishes. Postassessment process improvements need to be driven by senior management in order to achieve success. The key to successfully implementing assessment recommendations is to develop an actionable post assessment improvement plan. The plan should include several key components such as Process Improvement goals and objectives, project activities and schedules, team roles and responsibilities, and any risks or constraints. The organization's leadership team must take responsibility for leading the effort, vest real authority in those who are responsible for implementing the improvements, and keep stakeholders informed of the progress of the implementation process.

Chapter Summary

Chapter 3 presented the standard for assessing the capabilities of organizations that develop process-intensive systems. It also outlined a framework that process-focused organizations can use to better assess and improve organizational process maturity. It outlined the various levels of maturity and described several signs of process immaturity. From this chapter, we also learned that

- The Process Maturity Model contains a series of stages based on other well-known maturity models, such as the Software Engineering Institute's Capability Maturity Model.

- Each level describes how well the behaviors, practices, and processes of an organization can reliably and sustainably produce required outcomes.

- The Process Maturity Model can assist organizations deploy Process Improvement frameworks, evaluate organizational capability, conduct organizational benchmarking, identify appropriate areas for improvement, and elect high-priority improvement actions.

- The five levels of the Process Maturity Model are
  - Level 1, Informal
  - Level 2, Documented
  - Level 3, Integrated
  - Level 4, Managed
  - Level 5, Optimized

- Process Maturity Assessments are a diagnostic tool used to appraise and characterize an organization's processes.

- The scope of a Process Maturity appraisal can be any portion of an organization, such as the entire company, a specific business unit, a specific geographic region, or a specific project.

- An organization is ready to conduct a process maturity assessment when it becomes dissatisfied with any aspect of its current performance.

- Postassessment process improvements need to be driven by leadership in order to achieve success, and the key to successfully implementing assessment recommendations is to develop an actionable postassessment improvement plan.

- The primary difference between an evaluation and an assessment is that an evaluation is an appraisal where an external group comes in and looks at an organization's process capability to help make a decision regarding future business dealings, whereas an assessment is an appraisal that an organization performs to and for its own operations.

- The amount of time required to conduct a formal Process Maturity Assessment varies depending on the organization's size, complexity, and culture.

Chapter Preview

Now that the components of Process Maturity have been discussed, the next step is to learn the mechanics of Process Architecture, its benefits, and the tasks typically involved with planning and controlling all aspects of process creation, improvement, and retirement within an organization. Chapter 4 provides an introduction to the concept of a Process Ecosystem and illustrates how process architecture can help guide organizations toward higher process maturity.

CHAPTER 4

Process-Oriented
Architecture (POA)

Process Architecture is probably the least understood aspect of Process Improvement. However, the payoffs from a well-designed and well-managed architectural framework are enormous. Since processes adapt and change over time and are required to respond differently to varying environmental situations and strategies, this chapter provides an architectural model for planning, designing, and evaluating proposed changes in an organization's operating environment. The *Process-Oriented Architecture (POA)* approach is concerned with planning and controlling all aspects of process creation, improvement, and retirement within an organization. The development of an effective architecture framework to manage process-related changes is key to the success of any process-focused organization. Furthermore, a truly effective POA coordinates efforts across department boundaries to ensure effective and cost-conscious improvement efforts throughout all areas of an enterprise. As Process Improvement is not a one-time effort, processes need to continuously adapt and respond to changes in a company's environment, strategy, customer demands, business problems, and opportunities. The POA guides professionals as they propose changes to processes and related process components to respond to those events.

This chapter is organized around the following knowledge areas:

- *Enterprise Architecture defined:* What is the definition of Enterprise Architecture and how is Process Architecture related?

- *POA defined:* What is POA and how does it affect a company's Process Improvement efforts?

- *Core principles of POA:* What are the driving principles behind a POA approach?

- *Benefits of POA:* What are the key benefits to introducing and implementing a POA Framework?

- *POA Road Map:* What are the key POA components and how do they assist with managing a company's Process Improvement efforts?

Enterprise Architecture Defined

As business product and service cycles continue to change more and more rapidly, only flexible processes and maneuverable technologies can enable organizations to make the commitments required to continuously adapt and meet customers' needs. *Enterprise architecture* is the act of translating business vision and strategy into an effective enterprise change by creating, communicating, and improving the key requirements, principles, and models that describe the enterprise's current operations and enable its evolution. It is, first and foremost, an architecture approach used to describe the overall structure of an enterprise or a business unit. Second, it is about transforming a business and assisting organizations in making better decisions. This is accomplished when an organization gains an understanding of how complex their structures and processes are and ensures that proper business architecture and change management principles exist. Primarily, Enterprise Architecture seeks to model the relationships between the business, its processes, and any related components in such a way that key dependencies and redundancies from the organization's underlying operations are exposed.

The term *enterprise* is used because it is generally applicable in many circumstances including

- An entire business or corporation
- A subset of a larger enterprise such as a department
- A conglomerate of several organizations, such as a joint venture or partnership

The term enterprise in the context of Process Improvement includes the entire complex, sociotechnical system within an organization including

- People
- Information
- Technology
- Processes

NOTE *A common misconstruction is the notion that Business or Process Architecture is different than Enterprise Architecture, when, in fact, Enterprise Architecture is applicable to the entire organization including business, workforce, information, technology, and environmental perspectives. The belief is that Process Improvement professionals lead all process design and improvement activities in a way that encompasses all organizational viewpoints. At times, practitioners may focus more heavily upon one viewpoint or another, but they always link the interdependencies and interrelationships of these viewpoints to the business context and its temporal states (past, current, transitional, optional future, future, and retired).*

Process-Oriented Architecture Defined

Process-Oriented Architecture (POA) describes the philosophical approach to process business architecture, development, and management. It is a framework that contains a set of core elements, such as operating principles, benefits, deployment activities, and various related components, that are used to design and develop processes that are interoperable and reusable.

It is a method of Enterprise Architecture that focuses on the relationships of all structures, processes, activities, information, people, goals, and other resources of an organization including any technical structures and processes as well as business structures and processes. Process Improvement practitioners can use the POA to build all-encompassing Process Ecosystems, which demonstrate how processes, systems, and other process elements are aligned, and to assist with facilitating and expediting Process Improvement efforts throughout all phases of development.

The POA also oversees the maintenance of an organization's processes or capabilities and ensures that it depicts an organization both as it is today and as it is envisioned in the future. This includes both business-oriented perspectives and technical perspectives. In many ways, the POA serves as a communication bridge among senior business leaders, business operators, and Process Improvement professionals. Having this bridge ensures that all process-related improvements are coordinated and managed with an end-to-end perspective in mind rather than a traditional individualistic approach.

An effective POA is considered the foundation of process-focused organizations and enables enterprise visibility, alignment, and scalability. The diagram shown in Figure 4-1 illustrates the POA construct. In it, the linkage of processes to shared and reusable elements drives understanding of cause-and-effect relationships within an organization's entire Process Ecosystem, enabling better decision-making as processes are created or changed.

Core Principles of Process-Oriented Architecture

The Process-Oriented Architecture approach uses several architectural guiding principles to construct processes that business users can dynamically combine and compose into Process Ecosystems that meet continuously evolving and changing business requirements. The following guiding principles define the ground rules for development, maintenance, and usage of the POA to build Enterprise Process Ecosystems or business landscapes:

- *Loose Coupling:* This places emphasis on reducing dependencies among processes, procedures, and their related components. The primary goal is to reduce the risk that a change made within one process will create unanticipated changes within other processes.

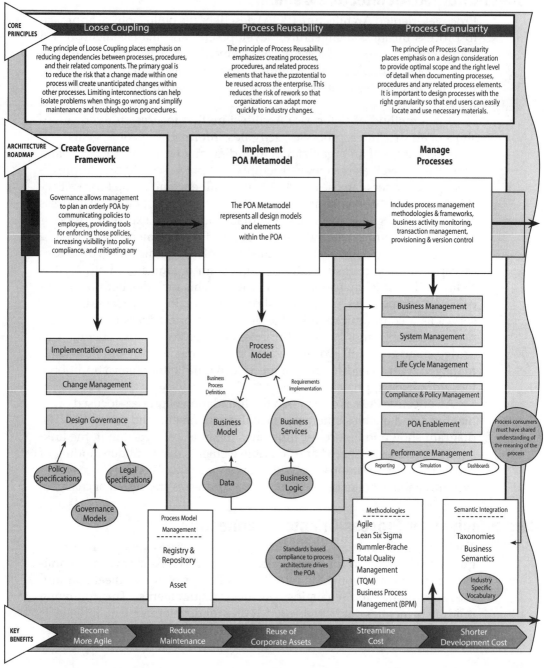

Process - Oriented Architecture (POA)

CORE PRINCIPLES

Loose Coupling

The principle of Loose Coupling places emphasis on reducing dependencies between processes, procedures, and their related components. The primary goal is to reduce the risk that a change made within one process will create unanticipated changes within other processes. Limiting interconnections can help isolate problems when things go wrong and simplify maintenance and troubleshooting procedures.

Process Reusability

The principle of Process Reusability emphasizes creating processes, procedures, and related process elements that have the pzzotential to be reused across the enterprise. This reduces the risk of rework so that organizations can adapt more quickly to industry changes.

Process Granularity

The principle of Process Granularity places emphasis on a design consideration to provide optimal scope and the right level of detail when documenting processes, procedures and any related process elements. It is important to design processes with the right granularity so that end users can easily locate and use necessary materials.

ARCHITECTURE ROADMAP

Create Governance Framework

Governance allows management to plan an orderly POA by communicating policies to employees, providing tools for enforcing those policies, increasing visibility into policy compliance, and mitigating any

Implementation Governance

Change Management

Design Governance

Policy Specifications

Legal Specifications

Governance Models

Implement POA Metamodel

The POA Metamodel represents all design models and elements within the POA

Business Process Definition

Requirements Implementation

Process Model

Business Model

Business Services

Data

Business Logic

Process Model Management
- - - - - - - - -
Registry & Repository

Asset

Standards based compliance to process architecture drives the POA

Manage Processes

Includes process management methodologies & frameworks, business activity monitoring, transaction management, provisioning & version control

Business Management

System Management

Life Cycle Management

Compliance & Policy Management

POA Enablement

Performance Management

Reporting | Simulation | Dashboards

Process consumers must have shared understanding of the meaning of the process

Methodologies
- - - - - - - - -
Agile
Lean Six Sigma
Rummler-Brache
Total Quality Management (TQM)
Business Process Management (BPM)

Semantic Integration
- - - - - - - - -
Taxonomies
Business Semantics

Industry Specific Vocabulary

KEY BENEFITS

Become More Agile | Reduce Maintenance | Reuse of Corporate Assets | Streamline Cost | Shorter Development Cost

FIGURE 4-1 Process-Oriented Architecture Construct

Process - Oriented Architecture (POA)

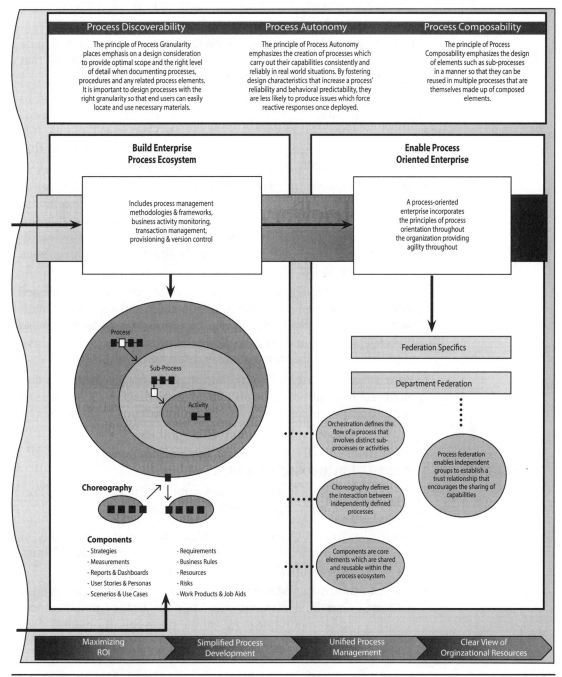

Process Discoverability

The principle of Process Granularity places emphasis on a design consideration to provide optimal scope and the right level of detail when documenting processes, procedures and any related process elements. It is important to design processes with the right granularity so that end users can easily locate and use necessary materials.

Process Autonomy

The principle of Process Autonomy emphasizes the creation of processes which carry out their capabilities consistently and reliably in real world situations. By fostering design characteristics that increase a process' reliability and behavioral predictability, they are less likely to produce issues which force reactive responses once deployed.

Process Composability

The principle of Process Composability emphasizes the design of elements such as sub-processes in a manner so that they can be reused in multiple processes that are themselves made up of composed elements.

Build Enterprise Process Ecosystem

Includes process management methodologies & frameworks, business activity monitoring, transaction management, provisioning & version control

Enable Process Oriented Enterprise

A process-oriented enterprise incorporates the principles of process orientation throughout the organization providing agility throughout

Process

Sub-Process

Activity

Choreography

Components
- Strategies
- Measurements
- Reports & Dashboards
- User Stories & Personas
- Scenerios & Use Cases
- Requirements
- Business Rules
- Resources
- Risks
- Work Products & Job Aids

Federation Specifics

Department Federation

Orchestration defines the flow of a process that involves distinct sub-processes or activities

Choreography defines the interaction between independently defined processes

Components are core elements which are shared and reusable within the process ecosystem

Process federation enables independent groups to establish a trust relationship that encourages the sharing of capabilities

Maximizing ROI

Simplified Process Development

Unified Process Management

Clear View of Orginzational Resources

FIGURE 4-1 *(Continued)*

Limiting interconnections can help isolate problems when things go wrong and simplify maintenance and troubleshooting procedures.

- *Process Reusability:* This emphasizes the creation of processes, procedures, and related process elements that have the potential to be reused across the enterprise. This reduces the risk of rework so that organizations can adapt more quickly to industry change.

- *Process Granularity:* This emphasizes the design consideration to provide optimal scope and the right level of detail when documenting processes, procedures, and related process elements. It is important to design processes with the right granularity so that end users can easily locate and use necessary materials.

- *Process Discoverability:* Here the emphasis is on the creation of processes that are discoverable by adding interpretable metadata to increase process reuse and decrease the chance of developing processes that overlap in function. By making processes easily discoverable, this design principle indirectly makes processes more interoperable.

- *Process Autonomy:* This places emphasis on the creation of processes that carry out their capabilities consistently and reliably in real-world situations. By fostering design characteristics that increase process reliability and behavioral predictability, they are less likely to produce issues that force reactive responses once deployed.

- *Process Composability:* Here, the emphasis is on the design of elements such as subprocesses in a manner so that they can be reused in multiple processes that are made up of composed elements.

Benefits of Process-Oriented Architecture

There are several benefits to using a POA approach when building Process Ecosystems within an organization. These include

- *Increased agility:* POA enables organizations to respond quickly to new business imperatives, develop distinctive new capabilities, and leverage existing processes for true responsiveness.

- *Reduced maintenance:* Maintaining a subprocess or process component as a contained element simplifies the task of maintenance. Changes can be made once and in one place for all business processes that use the component, reducing the effort required to make modifications.

- *Reuse of corporate assets:* POAs promote the reuse of existing assets, increasing efficiency and reducing process development and improvement costs.

- *Streamlined costs:* Introduction of POAs within an organization frees up resources and helps to ensure that investments are focused on core capabilities aimed at growing the business instead of on maintenance and modifications to multiple processes.

- *Shorter development times:* The major benefit of reuse is shorter process development times and reduced development costs. Development times are also compressed because they must conform to a consistent architecture, even though elements from one process can be used in another.

- *Maximizing return on investment on legacy processes:* As well as providing a level of reuse, joining a legacy process with newer elements created for other processes can prolong its life, allowing an organization to maximize its investment in legacy business processes that cannot easily be thrown away.

- *Simplified and productive process development:* A unified, easy-to-use set of process attributes and components enhances process productivity, promotes asset reuse, and fosters enterprise collaboration.

- *Unified process management and monitoring:* A POA that uses shared components and cross-process end-to-end tracking helps provide integrated governance and security.

- *Clear and comprehensive view of an organization's resources:* Relationships between people, processes, systems, and goals are easily understood, and changes can be quickly and easily analyzed and implemented.

Process-Oriented Architecture Road Map

Organizations may have several thousand processes with hundreds of subprocesses and related components. The POA road map guides organizations as they progress into a process-oriented operating model and helps them manage the growing number of operational processes that may exist across an enterprise. The POA focuses on delivering the greatest return on investment while ensuring proper process governance is in place. Practitioners can use the POA to build an all-encompassing Process Ecosystem that demonstrates how processes, systems, and other components are aligned. This aids in the identification of organizational enablers and processes that support business success and add value and those processes that do not.

The POA road map is comprised of five steps:

1. Create a POA governance framework.

2. Implement a POA process model.

3. Manage processes.

4. Build an Enterprise Process Ecosystem.

5. Enable the Process-Oriented Enterprise.

The POA assists with meeting the following business objectives:

- Developing and applying governance principles
- Determining how the enterprise environment supports the business strategy
- Identifying operating components that will add value to business processes
- Mapping existing systems, policies, procedures, and other components to business processes
- Setting Process Improvement priorities and assessing the impact of the new processes on the current Process Ecosystem

Creating a POA Governance Framework

Proper architectural governance is crucial when building all-encompassing enterprise Process Ecosystems. Governance ensures all process-related activities are carried out in a planned and authorized manner. This includes ensuring there is a sound business reason behind each process change; identifying the specific processes, people, and systems affected by a proposed change; and ensuring that all changes are executed based on the POA construct and guiding principles. For businesses that are required to comply with the Sarbanes Oxley Act, this is a critical business activity, as section 404 requires that any processes that could affect the company's bottom-line be auditable. This includes tracking who did what, to what systems, where and when, and ultimately why. Ensuring that an integrated POA governance framework exists enables organizations to implement end-to-end governance across the entire Process Improvement life cycle, thus simplifying how process authors and users discover, reuse, and change processes that effectively deliver on the promise of POA.

There are several subelements within the POA governance framework, including Design Governance, Implementation Governance, and Change Management and Operational Governance. These are discussed below.

Design Governance

Design Governance is an important part of an integrated POA framework that allows organizations to build process solutions according to plan, set priorities, model current and desired processes, and identify and prioritize candidate processes and process changes. Organizations often struggle with balancing long-term planning with the need to address immediate and short-term business requirements. As a result, the short-term requirements often take precedence over long-term strategic planning. When this is applied to enterprise process architecture, organizations often find themselves with several processes that deliver minimal business value and cause confusion and/or duplication.

Proper Design Governance for Process Improvement activities mitigates this issue and facilitates the identification, analysis, and modeling of candidate processes, policies, procedures, profiles, business rules, and other related process information. An effective Design Governance process examines existing processes and planned process improvements to determine which proposed changes should be deployed as new additions to a Process Ecosystem. It also registers reusable process elements to help speed value delivery wherever possible. In this phase of governance, management can assess the relevance and suitability of any proposed change to a Process Ecosystem quickly by using the POA as a reference architecture and source of control over the Process Ecosystem.

Design Governance also ensures that the right individuals within various organizations are coordinated to identify and analyze potential process improvements in a planned and managed fashion. This includes business owners and operators, enterprise architects, business analysts, project managers, process improvement managers, and portfolio managers who have an inherent stake in process work. When utilizing Design Governance mechanisms, processes can be proactively built and released into service as needed. This approach reduces late creation of single-purpose processes and promotes POA goals by avoiding chaotic process sprawl.

Being able to examine the impact of changes to the environment when modifying processes provides an essential service to the company. The more familiar team members are with the Process Ecosystem and its ongoing evolution, the more opportunities arise for improvements in its existing state of operation. The output from the Design Governance process is a set of candidate processes and changes that are inputs to the Implementation Governance process.

An example that illustrates Design Governance in practice is the deployment of a new product line within a company. When launching a new product, an organization must ensure that it can successfully design, build, and deploy the product into the market. Design Governance allows organizations to easily search for and reuse existing processes and components that support the organization's current infrastructure. In this case, the company is able to reuse and enhance its existing processes, such as its Order Processing process, with new features to handle the new product, rather than creating a new process and potentially duplicating time, effort, and capital.

Implementation Governance

Implementation Governance shepherds approved process changes and additions, as well as any related components, through the Process Improvement life cycle to ensure that the deployment of any proposed solutions is successful. Within Implementation Governance are methods to approve process deployment, ensure policy compliance, maintain adherence to applicable law, drive proper change management, analyze and assess for collusion, and provide clear segregation from other process changes.

In this phase, management can determine if process additions or improvements meet enterprise standards and guidelines before they are launched into an operational state. For example, for a process change to move into business execution, an organization may require that a series of accompanying artifacts and steps be completed in order to authorize the process improvement launch. This could include training and communication, process profiles, process maps, categorization and storage within the enterprise repository, and business endorsement from a process owner for the proposed process improvement. In many cases, Implementation Governance overlaps with Change Management and Operational Governance as process owners and process operators within operational environments respond to feedback about their performance, compliance with corporate strategies, consumer expectations, and feedback from internal team members regarding usability.

Change Management and Operational Governance

Change Management and Operational Governance ensure that processes are behaving correctly and provide a mechanism for absorbing change. When performance gaps are discovered, this oversight ensures that the proposed change actually closes performance gaps with POA guiding principles. Activities within this phase include process monitoring against performance targets and threshold tolerances, impact analysis, and strategic alignment assessment.

The Operational Governance process relies heavily on predefined business rules and business policies to ensure changes made within the Process Ecosystem do not disrupt ongoing business operations. A well-defined Change Management and Operational Governance process will fully detach process consumers and operators from the complexity of process change implementation and enforcement, change management standards, and process training and other nuances or impedances to interoperability. Similar to implementation governance, this phase also includes mechanisms to approve process deployment, ensure policy compliance, maintain adherence to applicable laws, drive proper change management, analyze and assess for collusion, and provide clear segregation from other process changes.

Often new processes are the representation of complex business rules. The mapping of proposed business rules on existing processes can reveal conflicts in business logic long before systems and people are expected to adhere to the new process. There are many cases in which two or more processes conflict and are only discovered postimplementation when proper inspection is not done during change management and operational governance reviews.

In a unified governance environment, processes governed within the Change Management and Operational Governance phase will have progressed through the Planning Governance and Implementation Governance phases. In many cases, organizations will have many existing processes that have not been subject to governance previously; however, it is recommended that they be integrated into POA using this governance framework.

Implementing Process Models

Establishing a common understanding of processes and subprocesses, associated business rules, and other related process components is an important milestone in establishing mature Process Ecosystems. Process Modeling is a technique used to enhance this activity and supports the effort of creating flexible process models that assist organizations in connecting critical process components across the business. The POA Modeling approach is a bottom-up method that consists of assembling integral atomic processes and/or subprocesses and components along with detailed descriptions into one central view. Its purpose is to document and communicate these components and enhance their reuse so that they are better managed and executed.

Process Modeling activity focuses on the practice of constructing process models that adhere to the guiding principles of the POA. These models provide visibility to the inner workings of a Process Ecosystem in order to make them improve and evolve as needed. They enable the development of more robust processes by providing a visual cue that can be used to understand the impact of performance and areas identified for change. As processes evolve, the associated Process Models that illustrate them must also change. New process models often have to be built or existing ones improved as more departments embrace a process-centric mind-set.

There are two key building blocks to implementing a proper POA Process Model:

- *Registry and repository:* Registries and repositories are built whenever information must be used consistently within an organization or group of organizations. Examples of these situations include organizations that

  - Transmit information through process- and/or procedure-level activities

  - Need consistent definitions of process-related information including processes, activities, organizations, and system names

- ○ Wish to ensure the accessibility of information across time zones, between databases, between organizations, and/or between processes

- ○ Are attempting to break down silos of information captured within processes and related applications

- ○ Suffer from poor collaboration and the communication that shared understanding of interdependencies would reduce

A Process Registry typically has the following characteristics:

- ○ Protected environment where only authorized individuals within an organization are able to steward changes (e.g., a Process Ecosystem Administrator)

- ○ Stored data elements that include both semantics and representations of the process components

- ○ Semantic areas of a Process Registry that contain the description of process elements with precise definitions

- ○ Processes that are represented in a specific format that all operators, users, and managers can understand

- ○ Process Model definitions and structures that have been reviewed and approved by appropriate parties

- *Proper asset management: A* Process Model requires a definitive library to assist with managing and governing the business and technology assets involved in process creation and improvement. An asset in this context is considered to be any published work product shared or referenced across the organization that meets a recurring business or technical need, is directly related to day-to-day activities executed by business users, and has a positive economic value to the enterprise.

  Assets are useful artifacts for process operators and improvement professionals that, in many cases, help facilitate or directly solve process-related issues if they can be found, used, reused, improved, measured, and governed. Ensuring proper asset management enables organizations to govern and reuse all types of assets, in turn, helping them optimize productivity and resources in their environment.

Regardless of the implementation, the imperative is to adhere to the POA standard of Modeling and link all processes together across the enterprise. It is the inter- and intradependencies between processes and their activities and data elements that help foster productive process improvement discussions. Process models are important as they provide the conduit to attach critical information such as where isolated use of corporate policy, law, compliance, business unit–level decisions/practices, requirements, business logic, work instructions, data sources, application dependencies, security attributes and logically see where each is used. They also act as the source for dynamically rendering points of consideration for effective governance. Consistency in approach and central management are paramount to the ability to harvest value from the investment in a Process Model. Models increase the productivity of process improvement managers and improve the quality of the models they produce, fostering reusability and extensibility within an enterprise Process Ecosystem.

Managing Processes

Process-Oriented Management refers to the practices recommended to keep Process Ecosystems in proper working order. In order to ensure proper maintenance and ongoing sustainability, the POA prescribes seven management practices: Business Management, Life-Cycle Management, POA Enablement, System Management, Compliance and Policy Management, Performance Management, and Semantic Integration. These are described below.

Business Management

Within every growing organization are activities that need to be performed to support process execution and governance. These activities include effective program management to wrap discipline around the quality-of-service targets the business sets out to achieve. This discipline ensures that operating goals are resident within the ecosystem of processes and threads those goals within each independent process, taking special care to reduce complexity and duplication wherever possible. In doing so, visibility of the entire process architecture is maintained. Productivity is increased and speed to value is improved in these cases.

Many organizations establish centers of excellence or departments with the sole purpose of governing their process landscape and its improvements. While this is more prevalent in large corporations where there are designated process officers or process improvement departments, the spirit of the intent can be achieved by grassroots cross-functional teams as well by adopting the constructs presented in this text. Regardless of the staff arrangement, the investment of time and attention toward disciplined management will return dividends. The most critical aspect of effective business management is to push information and decision-support data to flow across processes with transparency to everyone in the organization.

Life-Cycle Management

Process life-cycle management captures the following stages and phases involved in the conception, design, realization, and service of processes:

- *Conception: Imagine, specify, plan, and innovate. A process* starts to form as one imagines a way to accomplish an end goal and begins to outline a path to meet that goal. The conception phase is often iterative and rapid as it works to move to the next phase. Process conception presents an exciting opportunity to think without constraint about solving complex problems.

- *Design: Describe, define, develop, test, analyze, and validate.* After an idea is conceived, additional detail and consideration are given to known constraints and facts that result in a clearer design. In this step, information flows begin to be articulated. Simulation, assessment, validation, and optimization tasks are performed in an attempt to build as lean a process as possible with reuse in mind. Proper design at this stage also takes into account the core elements discussed in the Process-Oriented Core Principals: Process Reusability, Granularity, Discoverability, Autonomy, and Composability, as outlined in Figure 4-2.

- *Realization: Build and deliver.* Following the design phase is the realize phase in which the process moves through stages prior to being placed into service. This includes registration of the new process using the governance practices, review by an Architecture Review Board, and establishment of performance monitors/alerts/activities. Many organizations find automation of life-cycle management immensely helpful. There are many tools available that provide rapid prototyping in the design phase and act as a repository for processes such as iGrafx (www.igrafx.com), Aris (www.softwareag.com), and OpenText (www.metastorm.com). Once processes are manufactured, they can be checked against original requirements established in the design phase using computer-aided simulation software. In parallel to these tasks, training and change management

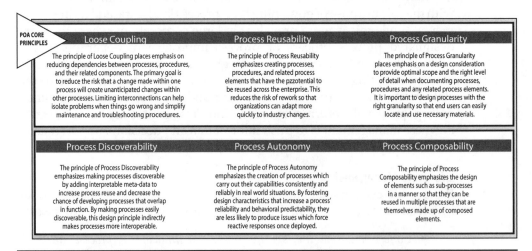

FIGURE 4-2 Process-Oriented Architecture Guiding Principles

documentation work takes place to prepare the organization for the process going into service.

- *Service: Use, operate, maintain, support, sustain, phase-out, retire, recycle, and dispose.* In the final phase of the life cycle, the focus is on management of service information. In this phase, information regarding the process is published for the user community along with support information needed to operate the process and understand its purpose and availability in order to extract maximum value. A process is maintained regularly in order to ensure that it is still achieving its purpose. Because every process comes to an end, unintended consequences of process retirement must be carefully considered. One benefit of following the POA is the transparency of interdependent processes, data, and outcomes. It is for this purpose that we emphasize the value of process traceability during each phase of process management.

Process-Oriented Architecture Enablement

Process-Oriented Management would not be complete without recognition of the value that enablement methodologies bring to the table. Agile, Total Quality Management, Lean Six Sigma, Business Process Reengineering, and Kaizen are well-documented methodologies that can be used to help organizations tune their processes to yield high quality. Each method is purpose built. Agile is engineered around delivery in an iterative fashion, while Lean Six Sigma eliminates waste from work/processes to achieve no greater than six defects per million opportunities. By extracting the best practices from methodologies such as these, organizations can better manage improvements at all stages of the Process Improvement life cycle.

Systems Management

Systems Management involves management of the information technology systems within the various processes in an enterprise. This includes gathering requirements, procuring equipment and software, distributing it to where it is to be used, configuring it, maintaining it, forming problem-handling processes, and determining whether objectives are being met. As processes become more and more reliable on technology solutions, the most critical element of Systems Management and Design is ensuring that process, not technology, is driving these requirements.

Compliance and Policy Management

Compliance and Policy Management is an organization's approach to mitigating risk and maintaining adherence to applicable legal and regulatory requirements. The POA framework recommends policy and compliance activities be integrated and aligned within an organization's Process Ecosystem in order to avoid conflicts and wasteful overlaps and gaps and to ensure proper visibility for all process users and managers.

A common business practice for Compliance and Policy Management is documenting business rules. Business rules are statements that define or control aspects of the business. They assert business structure and influence behavior by specifically describing the operations, definitions, and constraints that apply to an organization. By capturing and centralizing Business Rules within the Process Ecosystem, consistency of rules across process touch points is supported. This also increases visibility of operating practices and enables prompt and comprehensive decisions that comply with law, policy, contract-specific processes, and business unit policies.

The POA focuses on bringing business rules together with processes, subprocesses, and other components and merging them into an integrated, holistic, and organization-wide repository. In applying this approach, organizations can achieve such core benefits as ethically correct behavior, improved workforce efficiency and effectiveness, and improved compliance among the operators who regularly interact and use the processes.

Performance Management

Process-oriented performance management should encompass real-time availability and performance metrics of all processes within the Process Ecosystem, as well as design-time processes for comparison. Established key performance indicators that are attached to individual processes and process collection that aids in identification of processes that are both in and out of control create a competitive advantage. Effective performance management continuously compares the results of in-service processes with those retired and in design to ensure optimized processes are in service.

Semantic Integration

Process consumers and operators must have a shared understanding of the meaning of the various processes within their organization as well as any related components. A common and/or industry-specific vocabulary should be in place, one that includes taxonomies and business semantics for the terms used to describe process artifacts. A standard taxonomy gives stakeholders a common language to use when benchmarking, collaborating, and improving processes.

Building an Enterprise Process Ecosystem

Process Ecosystem is a term used to specify the processes in an organization that are directly involved in adding value to the company as well as their related components and supporting activities. These processes can be connected to each other in sequence, can be atomic, or can be composed of other subprocesses. Processes also contain several important attributes that can be added to form an all-encompassing repository of business information.

Enterprise ecosystems identify all the processes that are interconnected and driving toward business success. Two fundamental architecture concepts, in addition to creation, are used to build a Process Ecosystem within the POA construct: Orchestration and Choreography.

Process Orchestration is the coordination and arrangement of multiple processes or subprocesses exposed as a single aggregate process. Process Improvement professionals apply Process Orchestration to design and improve processes using existing process components within an organization. In other words, Process Orchestration is the combination of multiple process interactions to create higher-level business processes. With Orchestration, the process is always controlled from the perspective of one business organization.

Process Choreography is more collaborative in nature, whereby several processes are sequenced to run in connection with one another; these processes may involve multiple parties and multiple organizations. Process Improvement professionals utilize Process Choreography to design and improve processes in an organized manner across multiple organizations. With Choreography, the process is usually controlled from the perspective of multiple business units.

As illustrated in Figure 4-3, a Process Ecosystem should be thought of as multiple processes that are interconnected and related to their environmental conditions. Ever changing, adapting, and serving an individual and collective purpose, the ecosystem must also have sufficient maintenance to remain in good health.

The Enterprise Process Ecosystem is a clear recognition of existing interconnections between processes. Successful businesses recognize the choreography and orchestration needed to organize the essential activities within a process or subprocess, as well as the order and conditions that drive when to trigger a process to be initiated and completed. This forms dependent and interdependent understanding and, when placed into the POA framework, illuminates all players' roles within the ecosystem.

The ecosystem's health can be maintained by employing strategies that include having clarity about the purpose of each process; ensuring process requirements and intended value and the rules that govern them; determining what resources are engaged in supporting them; identifying the risks and thresholds that need to be in place in order to be responsive; using the scenarios and cases/situations that the process should be triggered on, as well as what subprocesses and procedures they spawn. Monitoring the Process Ecosystem requires putting in place measures that report targets and threshold zones as close to real time as possible. Often, performance dashboards aid process workers in understanding where bottlenecks are occurring. This enables intelligent inspection and resolution. Without the enterprise Process Ecosystem perspective, bottlenecks

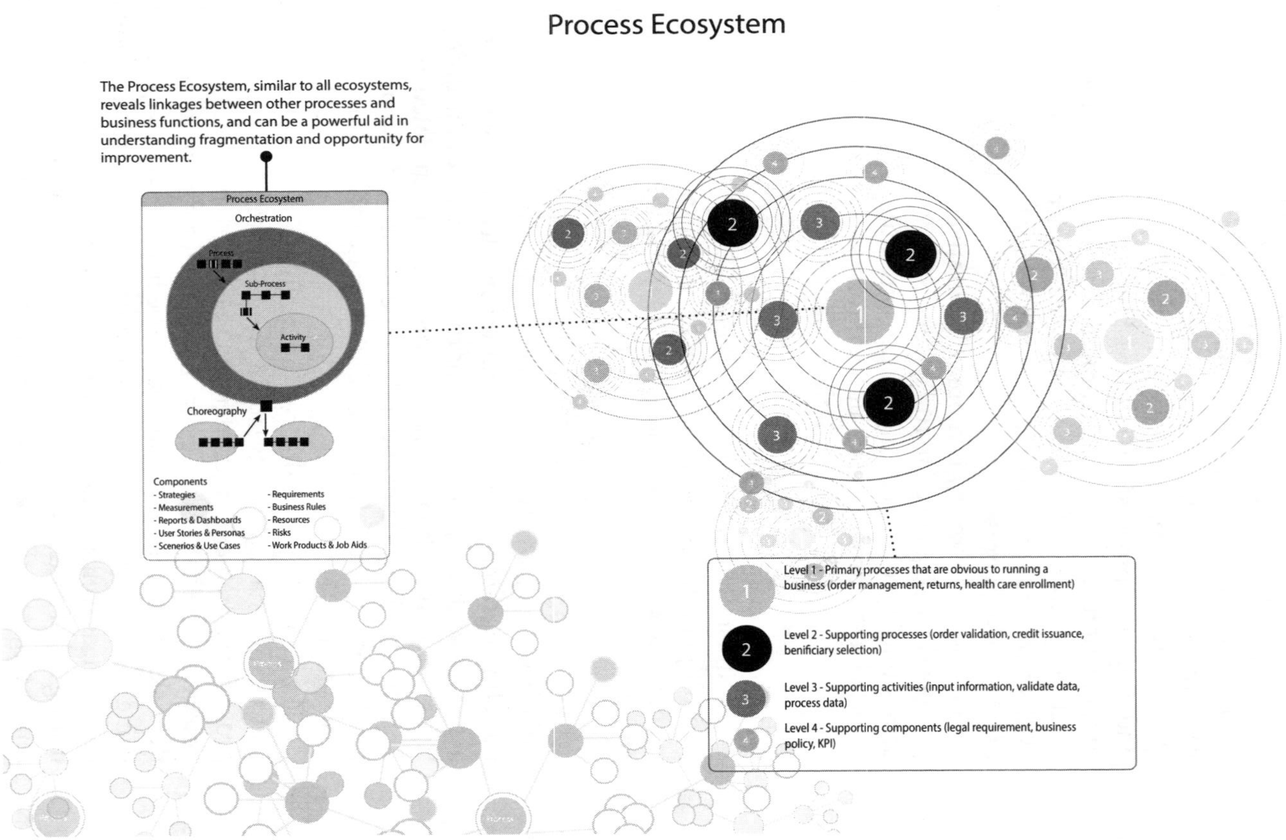
Process Ecosystem
The Process Ecosystem, similar to all ecosystems, reveals linkages between other processes and business functions, and can be a powerful aid in understanding fragmentation and opportunity for improvement.
Process Ecosystem
Orchestration
Process
Sub-Process
Activity
Choreography
Components
- Strategies
- Measurements
- Reports & Dashboards
- User Stories & Personas
- Scenerios & Use Cases
- Requirements
- Business Rules
- Resources
- Risks
- Work Products & Job Aids
Level 1 - Primary processes that are obvious to running a business (order management, returns, health care enrollment)
Level 2 - Supporting processes (order validation, credit issuance, benificiary selection)
Level 3 - Supporting activities (input information, validate data, process data)
Level 4 - Supporting components (legal requirement, business policy, KPI)

FIGURE 4-3 Process Ecosystem Overview

can be misunderstood and valuable time and resources can diminish the value of the project simply due to a lack of visibility of more serious problems elsewhere.

Enabling the Process-Oriented Enterprise

Every organization has processes that are used with varying degrees of efficiency within the context of a department or suborganization. A Process-Oriented Enterprise seeks to incorporate the extension of a standardized process across the entire organization and distribute, or *federate*, processes in a gapless fashion.

Process Federation is the capability to aggregate and leverage process-related information across multiple sources into a single view. This allows companies to reuse shared components. In essence, it is the act of federating, or joining, process components, departments, and users into a unified federation. This enables independent groups to establish a trusting relationship that encourages the sharing of capabilities.

Process Federation can be very challenging for many organizations. It requires significant understanding of an organization's entire business model, strong abstraction skills for those engaged in governance and process design, discipline, and conformity with the standards presented in the POA. It takes a high degree of discipline to achieve full federation. The complexity involved in designing processes for federation can be taxing too as the pressures of speed to value and individual departments challenge the value of federation to the core. This is especially apparent as the number of processes increases within the ecosystem. During these times, it is even more important to follow the framework in order to achieve the higher-order goals of a process-oriented enterprise.

Chapter Summary

In this chapter, we reviewed the definition of POA and its position as a core knowledge area within *The Process Improvement Handbook*. It provides a means to manage and understand change and describes how to use change as a strategic advantage and how to successfully install Process Ecosystems within an organization. Although it requires investment and maintenance, it yields a very definite return. From this chapter, we also identified the following key principles:

- There are six core principles of a POA: Loose Coupling, Process Reusability, Process Granularity, Process Discoverability, Process Autonomy, and Process Composability.

- POA enables organizations to respond quickly to new business imperatives, develop distinctive new capabilities, and leverage existing processes for true responsiveness.

- The POA road map guides organizations as they progress into a Process-Oriented operating model and is composed of the following five parts:
    - Creating a POA Governance Framework
    - Implementing a POA Process Model
    - Managing processes
    - Building an Enterprise Process Ecosystem
    - Enabling the Process-Oriented Enterprise
- Process Ecosystem is a term used to specify all processes in an organization as well as any related attributes that are directly involved in adding value to the company.
- Process Orchestration is the coordination and arrangement of multiple processes exposed as a single aggregate process.
- Process Choreography occurs when several processes are sequenced to run in connection with one another and may involve multiple parties and multiple organizations.
- Process Federation is the capability to aggregate and leverage process-related information across multiple sources into a single view.

Chapter Preview

We have covered the key terms used in Process Improvement environments, the importance of Process maturity, as well as the need to properly architect all processes within an organization. Chapter 5 discusses the use of the architecture principles found in the POA to construct an Enterprise Process Ecosystem.

CHAPTER 5

Creating a Process Ecosystem

This chapter describes how the principles of Process-Oriented Architecture (POA) are used to model and construct an organization's processes into a *Process Ecosystem.* The primary objective is to develop an overall view that integrates a company's various functional groups into one another in order to demonstrate how all activities within the organization are interconnected. A Process Ecosystem also links strategic goals to the necessary processes and helps coordinate the various Process Improvement efforts across an enterprise, including Information Technology (IT), Finance, Manufacturing, Supply Chain, and Marketing. In addition, it provides centralized visibility of any related process elements such as Business Rules, Key Performance Measures, as well as related system or technology details. If an organization's Process Ecosystem does not represent an integrated, cross-functional view, the business can fail to succeed in its Process Improvement efforts. This chapter provides an overview of the elements needed to orchestrate and choreograph an end-to-end Process Ecosystem.

This chapter is organized around the following topics:

- *Enterprise Modeling defined:* What is Enterprise Modeling?

- *Process Ecosystem defined:* What is a Process Ecosystem and how does it help organizations better manage their operations and Process Improvement efforts?

- *Process Ecosystem components:* What are the various components that form a Process Ecosystem?

- *Building a Process Ecosystem:* What are the fundamental activities involved in creating a Process Ecosystem and what does an Ecosystem model look like? What tools can be used to create and manage a Process Ecosystem?

- *Managing a Process Ecosystem:* How is a Process Ecosystem managed and who is responsible for administering it?

- *New Management obligations:* What are the critical issues in developing an effective Process Ecosystem and how does management's involvement in Process Improvement change?

Enterprise Modeling Defined

One of the principle reasons for transitioning from functional or silo management to true process management is to overcome issues related to performance, value, and communication loss between departments. To remain competitive, enterprises must become increasingly agile and integrated across their functions. However, simply changing functional titles into process roles does not fundamentally change an organization's landscape or provide the organization with an end-to-end view that employees can cooperatively manage and improve.

As we learned in Chapter 3, the practice of Enterprise Architecture is about transforming a business and assisting the business in making better decisions. This is done by understanding how complex the business structures and processes are and ensuring that proper process architecture and change management principles exist. We also learned that Enterprise Architecture seeks to model the relationships between the business, its processes, and any related components in such a way that key dependencies and redundancies are exposed from the organizations' underlying operations.

Enterprise Modeling is a component of Enterprise Architecture, and is a technique used to diagrammatically architect an organization's structure, processes, activities, information, people, goals, and other resources. Its primary purpose is to align company processes, resources, and systems with corporate goals and strategies to effectively enable compliance management, risk management, enterprise architecture, and Process Improvement efforts. An enterprise model can be both descriptive and definitional; it may cover both the current state and the "to-be" state of an enterprise's processes and is comprised of whole or smaller parts of an organization.

Process Ecosystem Defined

An ecosystem can be defined as a complex set of relationships within a systematic environment. It is, in essence, a community of objects together with their environment functioning as a unit. Interpreting an organization as an ecosystem of interconnected components can play a critical role in enabling more meaningful Process Improvement efforts and lead organizations, particularly those of significant size, to better process designs, better analysis of performance, as well as improved management of operations.

The term *Process Ecosystem* is used to describe the management of an enterprise as an integrated network in which all processes and related attributes are interconnected and driving toward business success. It provides an overview of the various processes within a company; their interdependencies; how information, products, or services flow in and out of each process; and stores this information in a centralized repository.

Business functions, processes, systems, applications, and other enterprise relationships need to be viewed as building blocks that can be reconfigured as needed to address changes in an organization's competitive landscape. By having an integrated and fully formed view of an organization, a Process Ecosystem can assist Process Architects and Process Improvement Managers in determining the impact of proposed changes or improvements on all parts of an enterprise. Practical examples include

- Determining how changing a process will affect resource consumption
- Determining how changing a process will affect activities of other processes
- Assessing whether changing a process will violate or break any business rules or regulatory requirements
- Evaluating whether changing a process will affect the quality of products or services provided by the enterprise
- Determining how implementing or purchasing a new machine or system will affect various processes
- Determining where training or communication will be required as new changes are introduced
- Assessing which processes may be affected by the introduction of a new strategic goal or policy
- Assessing which systems and applications require change as well as which services and data elements will be impacted by a change.

Process Ecosystems also assist process operators and stakeholders better understand how the company works as a whole and allows individuals to place their day-to-day activities within the context of the processes they are responsible for executing, the departments they work with, and the organization they are employed or affiliated with. When well-designed processes are arranged in a business process ecosystem, managers and staff can drill down through the model from the highest level of detail to the lowest level in order to view an assortment of organizational data. For example, an individual can search for where the function of Order Processing occurs within their organization, view the details of the Order Processing process, and click down to an individual activity within that process, such as entering an order into the company's financial system, and viewing its procedural-level detail. This level of interaction allows managers to measure how well processes, departments, and individuals are performing; adjust activities or even entire processes to improve performance; keep pace with changes in the business environment; or reflect new company strategies. Alignment at each level of the Process Ecosystem and across all parties involved helps ensure organizations remain agile and maintain their ability to quickly adapt to their customers' changing needs. Process Ecosystems also help enforce company-wide standardization and federation. Ultimately, building a Process Ecosystem means having the ability to view an organization as a cohesive whole in which the individual parts and the relationships between those parts can be identified and optimized. It is about identifying problems and opportunities and formulating changes to address those problems while exploiting Process Improvement opportunities.

NOTE *A Process Ecosystem is often referred to as a business landscape or process landscape. Each term carries the same meaning and possesses the same underlying purpose, that is, to better manage and improve business processes and related components in a cross-functional and centralized manner.*

Benefits of Modeling and Constructing a Process Ecosystem

Process Ecosystems help companies implement a structured approach to Process Improvement. Through a combination of POA principles and ecosystem modeling practices, organizations can align company processes, resources, and systems with corporate goals and strategies to effectively enable compliance management, risk management, and Process Improvement. Process Improvement Managers, Process Owners, as well as Process Operators and leaders can collaborate within one repository to build, share, and improve enterprise processes. Implementing a Process Ecosystem approach also helps those involved in Process Improvement projects that span several workgroups or divisions better share information.

A Process Ecosystem also

- Provides clear visibility into all of an organization's processes
- Provides greater understanding of the linkage and interdependencies among organizational components
- Enables end-to-end monitoring and correlation
- Identifies unused process or technology components that still consume resources
- Predicts future state process and technology needs to ensure sustainability
- Identifies gaps within process ownership and support
- Reduces the time, cost, and resource strain of Process Improvement activities
- Outlines any areas that require consolidation, retirement, or improvement

- Improves performance, transparency, and compliance
- Increases agility by easily analyzing the impact of process changes
- Helps evolve organizations from a silo functional design to cross-functional process design
- Increases fidelity and reuse of process assets
- Enables plug-and-play or "Lego-Brick" design of components into a composable enterprise model

Applicability of the Process Ecosystem

An ecosystem is a community of elements such as processes and subprocesses as well as smaller but related components that interact with these elements such as business rules and policies. Consequently, Process Ecosystems can vary greatly in size. For example, large Fortune 500 companies may be comprised of thousands of processes and hundreds of business rules, strategic goals, or technology systems, whereas small businesses may contain only a few dozen processes and related components. While several attributes may be found in both organizational sizes, the extent of the ecosystem and the quantity of items within may differ significantly. Additionally, organizations may elect to iteratively deploy ecosystem models across the enterprise in order to reduce the inherent risks with large transformation efforts. This means that an enterprise ecosystem can contain a series of smaller ecosystems that are linked together or operate as one large biodiverse system. Within any organization, large or small, disruptions or changes to processes can impact processes elsewhere with unforeseen results, unless they are visible and managed appropriately. In order to be truly agile, organizations require a degree of integration that is not possible without the use of an enterprise process model, and this is true for organizations in any industry and of any size.

Process Ecosystem Components

Process Components are the fundamental units or objects of a process. Process Ecosystem development uses an object-oriented or object-based approach to structure processes and associated attributes within an organization. It models an organization as a group of interacting objects where each object represents some unit of interest within the organization. Ultimately, structuring organizational processes into a Process Ecosystem means considering the enterprise, not as a set of separate functions that are performed, but as a set of related, interacting objects that work together to produce specific outcomes. Figure 5-1 outlines several common object groupings or categories. The task for Process Improvement professionals is to determine which objects or group of objects is being used by an organization (or should be) and to define each component,

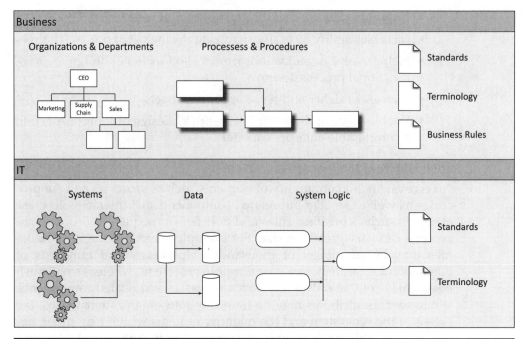

FIGURE 5-1   Sample Process Ecosystem Groupings

analyze and determine interrelationships among the various components, and remove any duplication or waste, ultimately driving the organization to higher maturity levels.

For example, in a manufacturing organization's Process Ecosystem, the manufacturing process is considered an object and the departments that execute the process are also considered objects. The Employees within each department are also objects as are any systems those employees use to execute the manufacturing process. Any policies or business rules that govern the way products are manufactured are also considered objects; however, they may also have specific relationships or associations with other processes outside of manufacturing. The most commonly used components that form an organization's Process Ecosystem are described in the following paragraphs.

Strategic Goals

Strategic Goals are planned objectives that an organization strives to achieve. They transform an organization's mission into specific performance targets that are measurable. Also, they outline what needs to be accomplished, when it needs to be completed, and who is responsible for its completion. Since strategic goals are often process-oriented or require changes to one or several processes, aligning a Process Ecosystem with a company's objectives helps stakeholders identify which processes will be affected by a particular goal and ensure that any required process improvement projects are focused appropriately.

Policies

Policies are basic principles and/or guidelines formulated in order to direct and limit actions in pursuit of an organization's Strategic Goals. Capturing an organization's policies and associating them with any relevant processes or related attributes ensures that anyone involved with operating or improving those processes is aware of any operating standards set forth by the organization's leaders, board members, and shareholders.

Processes

Processes are a series of activities or subprocesses taken to achieve a specific outcome. They are documented graphically by way of process maps and identify the sequence of activities needed to produce a particular outcome as well as illustrate who is responsible for each step. Documenting processes is the most critical component to building Process Ecosystems as processes are the backbone of process improvement efforts. Consolidating processes into a centralized Process Ecosystem repository allows people across the organization with a role in processes, such as Process Owners, Process Performers, or Process Improvement Managers, to have a transparent view into the part they play in the ecosystem and how any proposed changes might affect them.

NOTE *Subprocesses and activities are usually children of major, or parent, processes. These subprocesses and activities should also be thought of as separate and unique objects within an ecosystem as both may contain attributes or reusable elements that can be orchestrated or choreographed within other processes or areas in the business.*

Procedures

Procedures define the specific instructions necessary to perform the steps or activities in a Process. They define how process activities are performed and are typically documented in step-by-step fashion and describe in detail each activity within a process, including human- and system-driven tasks. Building procedure-level detail into the Process Ecosystem ensures that it will serve as a useful tool for all members of an organization. It enables Process Improvement Managers and project stakeholders to easily see workforce elements that will be affected by process changes and focus on how those changes might affect the Human Performance element. Also, they allow professionals to pinpoint just how proposed technology changes might affect how individuals or departments carry out their process activities.

Key Performance Measures

Key Performance Measures (KPMs) are a set of indicators that are represented at key points along a process or group of processes. They help organizations understand how well their people, processes, and systems are performing in relation to any strategic goals, standards, service-level agreements, or required process outcomes. Ensuring KPMs are attached to all possible processes and activities within an organization's Process Ecosystem enables stakeholders to understand whether the organization is on track or not and makes it possible to better pinpoint exactly where critical performance errors such as waste are happening. Being able to pinpoint performance anomalies and their exact location within an organization expedites root-cause analysis (RCA) efforts and enhances process improvement activities quickly if performance is insufficient.

Process Dashboards and Scorecards

A *Process Dashboard* is a user interface comprised of graphical information (charts, gauges, and other visual indicators) that identifies current or historical trends of an organization's KPMs. Linking dashboards of KPMs to the exact processes they represent in a Process Ecosystem allows organizations to monitor process performance and proactively identify discrepancies and potential quality issues. Furthermore, using dashboards helps organizations link process and people together through collaboration, and end users can work cooperatively to increase efficiencies through the use of dashboards. Since KPMs can be assigned to any process, composing dashboards from the menu of KPMs already identified in the ecosystem allows process owners to build reports that suit any level of an organization. Executives can view reports compiled across numerous processes and at various levels of detail, and staff members can view their individual performances.

Personas

Personas are detailed descriptions or depictions of customers and users of a process. Personas are based on real people and real data that humanize and capture key attributes about archetypal process customers. This data might include customer goals, preferences, personalities, and motivations. Building Personas into an organization's processes helps owners, operators, and improvement managers make informed design decisions and, more importantly, increases empathy within process designs. The primary notion is that if an organization wishes to design effective processes, the processes must be designed on both the customer's and the operator's requirements. Figure 5-2 depicts what a persona might look like for a customer of a product returns process.

Resources

Resources are the components required to carry out the activities of a given process or series of processes. Resources can be people, equipment,

---

Product Returns Customer Persona

The following example is only a small portion of a larger persona developed by the Process Owner and Process Improvement team.

Samantha Benson – Persona: "Smart Shopper"

Samantha is a 35-year-old married mother of three children. She has a University degree in Business Administration and spends what little free time she has enjoying fictional books. She is focused and goal oriented and takes great pride in her ability to raise her family with quality values. She is comfortable using a computer to chat and keep up with friends but always prefers the telephone as hearing a person's voice is much more engaging. She enjoys her job as a full-time quality assurance specialist and believes very much that the root of all good companies is the customer experience.

FIGURE 5-2   Sample Persona of a Process Customer

systems, facilities, or anything required for the completion of a process step. By modeling organizational resources, Process Improvement professionals are able to understand who is responsible, accountable, informed, and consulted for different activities, including human resources directly involved in a processes operation or support resources such as applications, systems, and equipment. Mapping resources into a Process Ecosystem allows for identification of impactful relationships and the ability to plan for how Process Improvements can best support the business. Typical resources that are incorporated into Process Ecosystems include

- Departments
- Roles
- Staff members
- Business communications
- Systems
- Applications
- Application services
- Data
- Facilities
- Business events

NOTE *RACI is an acronym used to describe the participation of various roles in executing a business process and is useful in clarifying roles and responsibilities in cross-functional processes and Process Improvement Projects. The acronym RACI was derived from the four responsibility categories typically used: Responsible, Accountable, Consulted, and Informed.*

Risks

A *Risk* is any factor that may potentially interfere with or impact the successful execution of an organization's processes or operations. Linking risks with organizational processes enables process owners and stakeholders to manage and understand problems that could affect their business goals, processes, and related attributes. Recognizing potential problems allows Process Improvement professionals to proactively attempt to avoid them through proper action.

Business Rules

Business Rules are statements that define or control aspects of a business. They assert business structure and influence behavior by describing the definitions and constraints that apply to an organization. Capturing and centralizing business rules within a Process Ecosystem allow organizations to efficiently and consistently apply rules across its various operations and processes and ensures that all compliance initiatives are being met. Building rules into the Process Ecosystem also increases employee visibility into operating practices and provides context for operators of a process as to why they are required to perform actions in a certain manner o sequence. Easy access to business rules also enables prompt and comprehensive Process Improvement decisions that comply with the law, company policy, business unit policy, and contractual obligations. Figure 5-3 illustrates several common business rule attributes that are attached to rules stored within a Business Process Ecosystem.

Other Components

Virtually any business element or artifact can be a component within an enterprise Process Ecosystem if it presents value to the organization and the way it conducts its operations, drives its processes, or improves its Processes and procedures. Other components that are often built into Process Ecosystems and utilized by Process Improvement teams include

- Process profile details (e.g., name, description, owner, suppliers, inputs, process, outputs, and customers (SIPOC), start point, end point)
- Job aides
- Data models and flows
- Organization charts

ID	Name	External Governance	Internal Policy/ Process	Description	Status	Status Reason	Hierarchy	Owner	Process ID	Process Name	Next Review Date	Previous Review Date	Effective Date	Expiration Date	Last Modified Date	Last Modified By	Created Date
1	Valid Account	SOX	N/A	All customers must have a valid customer account	Active	N/A	Law	Finance	AA-001	HHRP Process	01-15-2014	01-15-2013	05-01-2007	N/A	02-01-2013	John Smith	04-19-2007

Figure 5-3 Common Business Rule Attributes

- Projects
- User stories
- Use cases
- System logic
- Products
- Product categories
- Competitors
- Contracts
- Partners
- Financial information such as capital and operating expenditures

Building a Process Ecosystem

An organization's level of process maturity often characterizes the amount of knowledge it has about its own business processes. If the organization has a high level of process maturity, its processes can be modeled into an integrated ecosystem relatively easily. If the level of process maturity is low within an organization, meaning processes have not been identified or documented, the level of effort increases as significant cultural and structural changes are needed in order to formally change the enterprise to an integrated entity. In many organizations today, the task of shifting stakeholders into viewing the organization as a systematic environment and gaining buy-in to build a centralized view often fails. The primary reason for this is that organizations and practitioners do not take the holistic approach required to see the change through from conception to deployment. Implementing a Process Ecosystem and shifting an organization's mind-set to be process oriented can be a challenging endeavor. However, Process Improvement professionals can avoid this outcome by employing a step-by-step approach that can assist organizations through this effort. The typical steps involved with successfully building an enterprise Process Ecosystem are described in the following paragraphs.

Design the Ecosystem

The first phase involves designing the Ecosystem.

1. *Develop a vision:* Develop a vision to help direct both modeling and cultural change efforts along with a set of strategies for achieving that vision. This will help stakeholders see the need for change and the importance of acting immediately.

2. *Assemble a team:* Assemble a group of leaders and individual contributors with enough influence and authority to guide and lead the effort. Encourage risk-taking and nontraditional ideas, activities,

and actions. In many cases, the Process Improvement Organization takes the lead in building an organization's Process Ecosystem.

3. *Determine scope:* Identify requirements and the level of detail necessary for the Process Ecosystem. Determine the approach for modeling and mapping and what areas of the business will be included (e.g., Will it be the entire enterprise or only a subset? Will it be an iterative or waterfall deployment?). Last, determine what process components will be included in mapping efforts.

4. *Identify constraints:* Determine any expected roadblocks, limitations, or legal restrictions that enterprise areas might present, such as legacy processes or architectures, potential multisite issues, access restrictions, or data storage limitations.

5. *Select the proper tools:* Identify and analyze various industry tools that can be used to map and model the enterprise's processes and components. Select a tool that is simple to use and provides a positive experience for business customers (e.g., processes should be easy to find). Ensure that the tool is flexible and maintainable as process architectures should be easy to update.

Implement the Ecosystem

The next phase involves implementation of the Ecosystem.

6. *Catalog processes:* Determine which organizational processes have previously been documented and which processes require assembly. Establish process groupings and develop the contextual view of the organization. Also, determine if process components are widely understood, documented, and referenced.

7. *Standardize artifacts:* Ensure that common templates, terminology, and definitions are available for use and that all stakeholders involved in mapping efforts experience a seamless experience from Process Improvement Managers.

8. *Map processes:* Model and map all processes within the organization and collect and catalogue related components. Follow any architecture principles and change control methods as outlined in the POA framework.

9. *Add components:* Complement processes with relevant process elements such as associated business rules, measures, and procedures. Components can be documented throughout facilitation and mapping of the processes, or, in many cases, organizations can decide to iteratively enhance process details after initial mapping is complete.

10. *Deploy tools:* Launch a navigation tool for business users so that they can explore all enterprise processes, monitor performance, and subsequently participate in improvement efforts with a better understanding of process structure, reasoning, and performance.

Monitor the Ecosystem

The final phase involves monitoring the Ecosystem.

11. *Engrain processes into the culture:* Articulate the connections that exist between the new behaviors and the Process Ecosystem and develop the means to ensure ongoing usage and sustainability. Integrate process architecture and ecosystem practices into standard Process Improvement activities. Launch a formal Process Improvement training program so that process stakeholders understand their roles in operating and maintaining the Process Ecosystem as well as where and how to navigate organizational processes.

12. *Continuously analyze:* Analyze the new process and changes to ensure the ecosystem's architecture is maintained. Investigate structural alternatives and use increased credibility to propose improvement projects for changing processes, systems, structures, and policies. Ensure all changes utilize the principles found in the POA.

13. *Continuously improve:* Begin implementing process changes and continuously improve process performance. Complete any needed organizational structure changes to ensure proper support for ongoing process performance.

Development of a Process Ecosystem is neither the beginning nor the end of the Process Improvement journey. It is undoubtedly integral to the overall transition to process management as it focuses on what the organization is currently doing and how all of its parts are connected. It also fosters better change management but it does not replace traditional Process Improvement projects and activities, it simply better enables them.

NOTE *Process Mapping and Process Modeling are often described as two separate terms, where the activity of process mapping defines what a business does and who is responsible, and process modeling focuses more on the optimization of business processes. Both concepts create a graphical representation of activities within an organization in order to improve business processes, however, mapping is usually a one-dimensional exercise to graphically display a singular process. Modeling processes usually involves additional mapping of related processes and objects, to give an all-encompassing view of interconnected enterprise processes, activities, architectures, and components.*

Sample Process Ecosystem

As previously outlined, a Process Ecosystem is created by modeling and mapping the complex set of relationships that exist between an organization's resources, departments, processes, and related attributes. As such, Process Ecosystems are often likened to the human anatomy, whereby any critical function or functions of the enterprise (core processes) are articulated via modeling them as a backbone and any other components that support the organization's vitality (management or supporting processes) are modeled as extensions of the backbone and form the organization's skeleton or central nervous system. Figure 5-4 illustrates a sample Process Ecosystem designed with this comparison in mind.

This conceptual model outlines all of the core processes and related subprocesses within a mock organization. The organization's core processes are characterized through the center of the diagram and represent the organization's primary function at its highest level, including all of the processes required to successfully deliver outcomes to customers. Subprocesses are then laid out as children to their higher-order associations, and other processes such as organizational or support processes are listed within the model as attaching nodes or spokes. Process components are also attached to the various process areas, completing the makeup of the organization's Process Ecosystem at a macro level.

NOTE *Another common analogy for a Process Ecosystem is comparing it to a subway or transit system.*

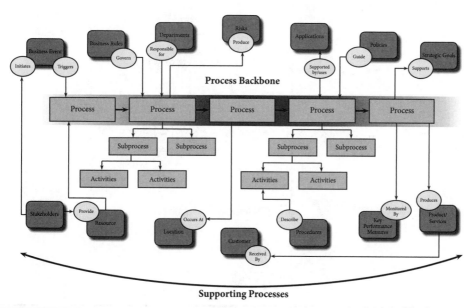

FIGURE 5-4 Process Ecosystem Concept

An organization's process components can also be represented in detailed form through hierarchies or folder structures that outline the various objects and classes of an organization's ecosystem. Figure 5-5 outlines a simple folder view of an organization's ecosystem components. With most of today's modeling and mapping tools, each

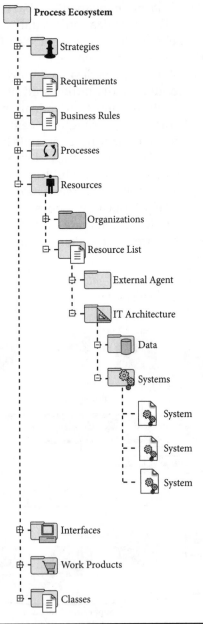

FIGURE 5-5 Process Ecosystem Structure

component is stored in clickable fashion so that anyone in the business can see a holistic view of each object, including attributes and details such as name, description, and relationships to other objects.

Process Backbone

As noted, the Process Ecosystem leverages the mental model of a Process Backbone. A Process Backbone is the set of core processes that run a business, and that all other processes and organizational components service. These core processes are structured to optimally deliver value as an interlinked process chain. The Process Backbone provides the ability to monitor the performance measures at a high-level collection point that signals business process health. Additionally, the process backbone aids in the identification of cost points that can be inspected for opportunities to reduce expense and address quality issues and wastes within the context of the end-to-end business model. As the enterprise collection of processes becomes more mature and the holistic landscape takes shape, the Process Backbone can be used to simplify understanding, provide executive summary, and generate critical attention to areas that need improvement within an organization. Ultimately, a Process Backbone represents an organization's core function at its highest level, illustrating all of the components required to successfully deliver value to customers.

Process Modeling and Management Systems

Process Modeling, Analysis, and Improvement systems continue to grow in importance as Business Managers, Process Architects, and Process Improvement Managers seek to better understand, streamline, and automate business processes. Process Architects use these tools to work collaboratively with business stakeholders to build out an organization's process architecture and to ensure that all processes conform to the requirements, principles, and models of the organization's change agenda. These tools are also used by Process Improvement Managers and Process Stakeholders to capture the details of business processes, ensure that process flows are properly mapped, and ensure that the quality of as-is and to-be gap analyses in Process Improvement projects are improved. Furthermore, these tools can serve as a bridge to improve the alignment of various efforts across an organization such as IT, Facilities, or Infrastructure initiatives. Process modeling as part of Process Improvement is becoming a starting point for a growing number of process management and compliance projects. Process modeling can help combat the common pitfall of organizations rushing to execution without truly gaining an understanding of root problems or causes. Equally, Process Improvement and Management Systems make it possible to simulate possible changes and improvements to a process or series of processes in an organization's Process Ecosystem. There are several solutions in the marketplace today that can assist organizations model their Process Ecosystems as well as help optimize business processes. These include

- iGrafx® Enterprise Modeling
- ARIS by Software AG
- Metastorm by OpenText
- Process Manager by IBM
- PowerDesigner by Sybase
- MEGA Suite by Mega International

NOTE *Many Process Improvement solutions can be used by enterprise architects as part of the Enterprise Architecture tool market. Given how closely linked the activities of architecture, modeling, and Process Improvement have become, several tool sets outlined in this section are now focused on providing all-encompassing tools for the market. These tools can be used to build out the architecture of an organization's Process Ecosystem, model and map any processes and process components, execute Process Improvement projects, and simulate the effects of proposed changes.*

Benefits of Using Process Management Systems

There are several benefits of investing in modeling and improvement tool sets, including

- The ability to organize, manage, monitor, and deploy all process artifacts, applications, and services from a scalable central repository and control center
- Increased leadership's confidence that process changes align with strategic endeavors
- Full business user visibility, making it easy for process owners and stakeholders to engage directly with Process Improvement efforts
- Process versioning and change management capability
- Federated view of process tasks across the enterprise
- Ability to integrate Process Improvement Projects with all process elements in an organization
- Dashboards that provide visibility into process performance and enable process stakeholders to respond to events in real time
- Ability to search and share content between process operators
- Elimination of static collections of Excel, Visio, and PowerPoint documents
- Easier Process Mapping through the use of templates and customized shape pallets

Simulation and Optimization

The act of simulating processes can be defined as imitation of the operation of a real-world process over a specific period of time. Once a Process Ecosystem is developed and all processes are documented, including all manual and system activities, a significant benefit of Process Management tools is that they allow Process Improvement practitioners to simulate the effects of process changes and analyze different scenarios in order to solve process performance issues. A common issue that arises throughout Process Improvement efforts is that a process may be fully capable of delivering a desired output but it may not do so at an acceptable performance level. In other words, a newly designed process may be functional but may not be optimal. *Process Simulation* enables all stakeholders involved in an improvement initiative to identify where process bottlenecks may occur in newly designed processes and where changes will benefit performance the most. Some examples of how Process Simulation can benefit Process Improvement Projects include

- Predicting how long process activities will take compared to an existing process design
- Predicting how often a process may be initiated
- Predicting the cycle time of each process activity
- Determining where resource constraints may impact performance

---

NOTE *Process Simulation is a subset of process modeling and mapping. It uses preset information such as performance thresholds and KPMs along with other data elements that are added to a Process Ecosystem to predict the outcome of various scenarios and proposed changes to processes within an organization.*

Managing a Process Ecosystem

As we've learned, Process Ecosystems help organizations gain an overview of their main processes and their interdependencies. However, the creation, improvement, and ongoing monitoring of an organization and its processes, particularly an organization of significant size, are not simple endeavors. After an organization produces a robust model on an enterprise scale, the model requires ongoing maintenance. For this reason, Process Improvement Organizations often generate two roles for managing the construction and ongoing sustainability of their Process Ecosystem and accompanying tools: the Process Improvement Architect and the Process Improvement Coordinator.

Role of a Process Improvement Architect

A Process Architect typically architects and designs an organization's processes. Process Architects are responsible for executing and maintaining

POA principles and building the organizational ecosystem model needed to support the organization's processes and any associated components. They also work to resolve any differences between Process Improvement Managers and process owners from within the business units or organizations they support. The Process Architect's role involves documenting the interrelationships between processes, people, and systems and identifying the hierarchy of processes, subprocesses, and process attributes that make up the Process Ecosystem. The Process Architect's overall aim is to ensure that all process attributes are managed and organized in one central platform and that all improvement efforts meet the organization's overall business strategy and do not duplicate or break other process components. Other responsibilities include defining and managing all ecosystem objects, determining interrelationships between objects, ensuring enterprise objects are reused across multiple models, guiding process development and improvement, and modeling and managing the organization's hierarchies. Ultimately, the Process Architect's primary goal is to uphold the principles found in the POA in order to increase the speed of development for Process Improvement efforts and ensure the sustainability of the enterprise's operations and processes.

Role of a Process Improvement Coordinator

The Process Improvement Coordinator is generally responsible for maintaining any procedural and administrative aspects of the Process Improvement Organization and serves as the designated point of contact for all Process Ecosystem issues and questions. The Project Coordinator's primary responsibility is the maintenance and housekeeping work involved in keeping the Process Ecosystem and any Improvement Projects assigned to the organization running smoothly. This includes a variety of tasks such as ensuring that repository permissions are in place, changes to the ecosystem are communicated, any standards that are required for its use are documented and distributed and administering the overall solution.

The Process Improvement Coordinator's overall role is to serve as the Process Improvement Organization's primary point of contact for the Process Ecosystem and accompanying tool kit. The coordinator is publicly listed as the individual to whom questions concerning the Process Improvement Organization and the Process Ecosystem can be directed. In addition, the Process Coordinator leads the drafting of Process Improvement guidelines, coordinates the implementation of ecosystem changes, and may also be required to facilitate small Process Improvement initiatives and change requests.

New Management Obligations

Managing complex work is never easy, and Process Improvement is a continuous cycle of complex improvements and change management efforts. Along with managing these complex changes, leaders in Process-Focused enterprises must also extrapolate the tools and techniques of Process Improvement to change how people, technology, and processes are managed within the context of the work environment and new leadership paradigms to deliver better results. Traditional Process Improvement focuses primarily on the improvement of one process at a time, the organization of the project team, the project artifacts, and the rules under which the project team operates. Managing through a Process Ecosystem requires leadership to think about the entire work environment as well as the specific leadership abilities expected from the organization's managers and stakeholders. Being Process focused is a fundamental paradigm shift, with the objective of making the organization more successful and the people within the enterprise happier in order to deliver better results.

Getting managers to become more Project focused and to think in terms of what is best for the overall organization requires changing behaviors and using a more democratic approach to management. More specifically, Leadership in process-focused organizations requires managers to

- Empower management, process owners, and team members through self-organization and commitment to results

- Transfer authority to process owners and stakeholders in order to determine how best to accomplish their tasks

- Transfer decision-making to individuals who are closest to the activities

- Demonstrate a greater openness to ideas and innovations

- Provide the necessary support and resources to Process Design and Improvement activities so that they successfully accomplish their expected results

- Become a change agent within the organization by accepting and publicly endorsing the idea that the status quo is not acceptable and that the old methods are no longer adapted to the organization's new reality

- Systematically involve business people in the definition and execution of solutions

- Adapt the style of management so as to use an inclusive and democratic approach

Chapter Summary

As we've learned, the dynamic nature of today's markets and the rapid introduction of disruptive technologies place a huge burden on businesses. Today, more than ever, businesses must be agile in strategy, innovative in technology, and responsive to customer demands. In order to do this, businesses must be able to make decisions faster. In addition they must improve required business processes rapidly. Static and siloed approaches, isolated project efforts, and static collections of Excel, Visio, and PowerPoint documents are no longer sufficient in attaining business excellence. Enterprises must manage themselves as an integrated network in which people, processes, and IT are blended into in an end-to-end Enterprise Process Ecosystem. The result is lower costs, better quality, higher throughput, faster reaction times, and proactive error correction. In this chapter, we also learned the following key concepts:

- Enterprise Modeling, which is a component of Enterprise Architecture, is a technique used to diagrammatically architect the organization's structure, processes, activities, information, people, goals, and other resources.

- Process Ecosystem is a term used to describe the management of an enterprise as an integrated network in which all processes and related attributes are interconnected and driving toward business success.

- Process Ecosystem development uses an object-oriented or object-based approach to structure processes and associated attributes within an organization.

- There are 10 common Process Ecosystem components. These include Policies, Processes, Procedures, KPMs, Dashboards, Personas, Resources, Risks, and Business Rules.

- There are three phases to building a Process Ecosystem: Design, Implement, and Monitor.

- A Process Architect typically performs the task of architecting and designing a Process Ecosystem.

- The Process Improvement Coordinator is generally responsible for maintaining all of the procedural and administrative aspects of the Process Improvement Organization and an organization's Process Ecosystem.

Chapter Preview

At this stage, we've discussed the principles of Process Architecture, Process Modeling, and what constitutes a Process Ecosystem. Chapter 6 takes you through a basic Process Improvement framework that includes common Process Improvement methods and techniques so that you can select the right tool to achieve effective improvement within an organization.

CHAPTER 6

Managing Process Improvements

Understanding processes so that they can be improved requires the application of basic methods, practices, and approaches. These, in turn, require that an appropriate organizational structure exists that can guide improvement efforts and ensure that activities are managed in a meaningful and harmonious manner. This includes tools for individuals who operate processes, those who manage and execute Process Improvement initiatives, as well as leaders who guide the enterprise and prescribe strategic imperatives and policies. This chapter outlines the components that make up a Process Improvement framework and the relationship of Process Improvement to strategic planning efforts. It discusses several methods that are used to conduct Process Improvement and provides a concrete model for discussing, implementing, and assessing the practice of continuous Process Improvement within a team or organization to meet growing customer demands. This chapter is organized around the following ten topics:

- *Process Improvement Framework:* What is a framework? What are the various components that make up a comprehensive Process Improvement Framework?

- *Environmental Factors and Organizational Influences:* What internal and external environmental factors surround or influence Process Improvement efforts?

- *Organizational Profile:* What is an Organizational Profile? How can it assist with assessing an organization's environmental factors?

- *Leadership:* What role does leadership play in Process Improvement activities?

- *Strategic Planning:* What is the definition of Strategic Planning? How does Hoshin Planning assist with these efforts?

- *Process Management:* What are the various Process Improvement methodologies? When should each methodology be used?

- *Performance Management:* What is Performance Management and how does it benefit an organization? How is performance monitored and communicated?

- *Workforce Management:* What is Workforce Management? What tools can be leveraged to effectively manage an organization's people?

- *Quality Management:* What is Quality Management? What defines a quality-focused organization?

- *Knowledge Management:* How is Knowledge Management defined? What tools are used to realize the benefits of Knowledge Management?

Process Improvement Framework

Since Process Improvement requires simultaneous changes in the process system, technical system, behavioral system, and management system, implementing a Process Improvement philosophy is a continuing and long-term goal. It can deliver results quickly, but can also take longer periods of time before becoming a core aspect of an organization's culture. Implementing a cohesive Process Improvement framework is a means of guiding organizations toward truly engraining Process Improvement in all aspects of operations.

In general, a *framework* describes the rationale of how an organization creates, delivers, and captures value. It is a structure intended to guide or steer an organization toward building particular practices or philosophies into its culture and operations. It provides a broad overview or outline of interlinked items such as strategies, tactics, and methods that support a particular approach and serves as a guide that can be modified as required.

Ten categories are involved in building a culture of Continuous Improvement and define the Process Improvement framework. These categories are placed together to emphasize the importance of aligning organizational strategy with customers and environmental factors. They also illustrate how an organization's workforce and key processes accomplish the work that yields overall performance results. Each element is critical to the effective management of an organization that wishes to continually improve performance and competitiveness. Figure 6-1 depicts the ten basic elements of the Process Improvement Framework.

Within the framework, all actions point toward results, which is a composite of environmental (action 1), management (actions 2, 3, 4), and workforce (actions 5, 6, 7, 8, 9) results. The larger two-headed vertical arrow in the center of the framework links the management triad to the workforce quintet, which is the linkage most critical to organizational success. A management team that works closely with its workforce ensures that employees are engaged, aligned, and informed. The larger single vertical arrow indicates the central relationship between the workforce and results, as the workforce is paramount in achieving organizational goals. All other arrows in the framework are two-headed arrows that indicate the importance of continuous feedback in an effective performance management system.

Figure 6-1   Process Improvement Framework

Environmental Factors and Organizational Influences

Enterprise environmental factors refer to both internal and external factors that surround or influence an organization's, a department's, or a project's success. These influences may come from multiple categories or entities, can enhance or constrain improvement efforts, and may have a positive or negative influence on an organization's activities. Example environmental factors and organizational influences include

- *External driving forces:* Customer, Political, Economic, Social, Technological, Legal, Regulatory, Marketplace, Shareholder, Media, Infrastructure (e.g., government or industry standards, regulatory agencies, codes of conduct)

- *Internal driving forces:* Culture, Structure, Processes, Communication Channels, Resourcing (e.g., staffing guidelines, training and skills, knowledge, employee morale, facilities)

Despite efforts to assess any and all environmental influences, disasters, economic shifts, and other unforeseen events, these negative factors can affect an organization at any point in its strategic road map. To the extent that they can, organizational leaders should be observant of the potential for unexpected environmental impacts and put contingency plans in place to ensure the least possible negative impact on the organization's success.

Organizational Profile

An *Organizational Profile* provides a high-level overview of an organization's environmental factors and influences. It addresses an enterprise, a department, or a project; its key relationships; its competitive environmental and strategic context; and its approach to improving performance. It also provides critical insight into the key internal and external factors that shape an organization's operating environment. These factors, such as the corporate vision, mission, core competencies, values, strategic challenges, and competitive environment, impact the way an organization is run and the decisions it makes. As such, the Organizational Profile helps organizations better understand the context in which they operate; the key requirements for current and future business success; and the needs, opportunities, and constraints placed on the organization's management systems.

Super System Map

Among the most widely used tools for creating an organizational profile is the *Super System Map* developed by the Rummler–Brache Group. The Super System Map is a picture of an organization's relationship to its business environment and helps to understand, analyze, improve, and manage these relationships. Management teams use this diagram to perform systematic strategic reviews. This is done by working through all key components of the organization's operating environment in order to analyze its current state and predict future scenarios. It can help to make strategic thinking more visible and can communicate a common view of the enterprise to all employees and stakeholders.

The Super System Map includes the following six key components, which are discussed in the following paragraphs:

- Organization
- Customers
- Suppliers
- Competitors
- Environmental influences
- Shareholders/Stakeholders (if applicable)

In addition to each component listed above, the map shows the organization's relationship with these components. These are displayed as arrows that show the flow of key inputs and outputs, including products, services, money, people, regulations, and information. Figure 6-2 depicts a basic Super System Map Template.

Organization: This box addresses the key characteristics and relationships that shape an organization's environment. Often populated with

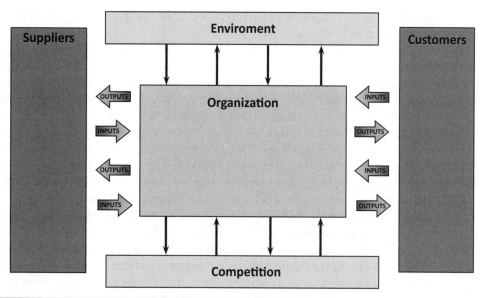

Figure 6-2   Super System Map

purpose, vision, mission, values, or core competencies, this section provides a clear understanding of the essence of the organization, why it
exists, and where senior leaders want to take it in the future. This clarity
enables leaders to make and implement strategic decisions that will affect
the organization's future and its processes. Examples of other information placed in this section include Organization Charts, Hoshin Kanri
information, and Business Plan information.

Suppliers: In many organizations, suppliers play a critical role in processes that are pertinent to running the business and to maintaining or
achieving a sustainable competitive advantage. There is often a significant or complex give-and-take relationship between a supplier and an
organization, and documenting these relationships and understanding
their complexities helps to ensure positive business outcomes. Suppliers
listed in this section may include staffing providers, Information Technology (IT) suppliers, professional service providers, supply chain partners,
outsourced services, research firms, and lending institutions.

Competitors: This box provides an understanding of who an organization's competitors are, how many there are, and their key characteristics.
It is essential to list and document competitors in order to determine an
organization's competitive advantage in the industry and marketplace.
Leading organizations have an in-depth understanding of their current
competitive environment, including the factors that affect day-to-day performance and factors that could impact future performance. Types of
competitors that may be listed in this section include Traditional (same

products, same markets, mature), New (new and growing), Emerging (disruptive technology, lurkers).

Environment: The environment in which an organization operates (regulatory or otherwise) places requirements on the organization and its processes and impacts how business is run. Understanding the organization's environment helps with making effective operational and strategic decisions. It also allows organizations to determine whether they are merely complying with the minimum requirements of applicable laws, regulations, and standards of practice or exceeding them. Environmental contributors include inputs from the Industry, the economy, and the general public.

Customers: This section lists any relevant customers of an organization. Knowledge about customers, customer groups, market segments, and potential customers allows organizations to tailor processes, develop a more customer-focused workforce culture, and ensure organizational sustainability. This section also addresses how the organization seeks to engage its customers, with a focus on meeting customers' needs, building relationships, and demonstrating the value of processes and services. Understanding the customer's voice provides meaningful information not only about a customer's view but also about a customer's marketplace behaviors and how these views and behaviors contribute to the organization's sustainability in the marketplace. Customer types may be listed by product type, user category, geography, type of buyer (individual, corporate, small business), distributor, and/or end-use customer.

Understanding an organization's profile, its strengths, vulnerabilities, and opportunities for improvement is essential to the success and sustainability of an organization and its processes. With this knowledge, organizations can identify their unique processes, competencies, and performance attributes, including those that set them apart from other organizations, those that help to sustain competitive advantage, and those that they must develop in order to sustain or build their market position.

Leadership

Leadership is defined as the art of motivating a group of people to act toward achieving a common goal. Leaders serve as the inspiration and directors of the actions within an organization and help pave the way for Process Improvement efforts. Senior leadership's central role in Process Improvement is to set values and directions, communicate goals and objectives, identify resources, and provide time and materials to execute improvement activities.

Responsibilities of senior leadership include

- Setting directions and creating a customer focus
- Setting clear and visible values for the organization

- Creating strategies and systems for achieving performance excellence
- Stimulating innovation and building knowledge and capabilities
- Ensuring organizational sustainability
- Inspiring the workforce to contribute and embrace change
- Ensuring appropriate governance bodies are in existence

Senior leaders should serve as role models through their ethical behavior and their personal involvement in planning, communicating, coaching the workforce, developing future leaders, reviewing organizational performance, and recognizing members of the workforce. As role models, they can reinforce ethics, values, and expectations while building leadership, commitment, and initiative throughout an organization. In highly respected organizations, senior leaders are committed to developing the organization's future and to recognizing and rewarding contributions by members of the workforce.

Strategic Planning

In most organizations today, some sort of structured planning is necessary to advance the organization's objectives. *Strategic Planning* is an organization's process of defining its strategy or direction and making decisions on allocation of its resources to pursue this strategy, including its capital and people. In essence, strategic planning is the formal consideration of an organization's future course and gathers senior leaders, managers, and any necessary process owners within the organizations in order to

- Determine key strengths, weaknesses, opportunities, and threats (SWOT)
    - *Strengths:* Characteristics of the organization that give it an advantage in the market
    - *Weaknesses:* Characteristics of the organization that put it at a disadvantage in the market
    - *Opportunities:* Items the organization could exploit to its advantage
    - *Threats:* Items in the businesses environment that could cause difficulty for the organization
- Determine core competencies and the ability to execute strategy
- Optimize the use of resources to ensure the availability of a skilled workforce
- Bridge short- and long-term requirements that require capital expenditure, technology development, or new partnerships
- Ensure that deployment of any vision will be effective

- Achieve alignment on all organizational levels: (1) the organization and executive level, (2) the key work system and work process level, and (3) the work unit and individual job level

- Ensure that organizational plans and requirements are communicated

- Ensure Process Improvement efforts are progressing and are using industry best practice

Process Improvement emphasizes the following three key aspects of strategic planning, which are critical to ensuring effective outcomes:

- *Customer Focus:* Focusing on the drivers of customer engagement, the customer experience, new markets, market share, key factors in competitiveness, profitability, and organizational sustainability.

- *Operational Performance Improvement Focus:* Focusing on short- and longer-term productivity growth and cost/price competitiveness. This focus helps to build operational capability, including speed, responsiveness, and flexibility and represents an investment in strengthening organizational fitness.

- *Organizational and Personal Learning Focus:* Focusing on the necessary strategic considerations in today's environment and the training and learning needs of the workforce, thereby ensuring that improvement and learning reinforce organizational priorities.

The primary role of strategic planning is to guide an organization's actions and decisions over a specific period of time. A strategic plan that is continually reviewed helps an organization's employees stay on track with tasks that are critical to long-term objectives. Many organizations are able to achieve this kind of routine, but many suffer from negative conditions that can prevent success. Common pitfalls organizations should steer clear of when conducting Strategic Planning efforts are

- Establishing too many priorities
- Not providing sufficient detail
- Not continually reviewing, planning, and prioritizing

Hoshin Kanri

Among the most widely used tools for strategic planning is a model known as *Hoshin Kanri,* also termed Policy Deployment or Hoshin Planning. Hoshin Kanri was devised to capture and cement strategic goals about an organization's future and what it means to bring these to fruition. Popularized in Japan in the late 1950s, *Hoshin* means compass or pointing the direction and *Kanri* means management or control. The name suggests how planning aligns an organization with accomplishing its goals. It is a system in which all employees participate, from the top down and from the bottom up, so that everyone in an organization is aware of their own and management's Critical Success Factors (CSFs) and Key Performance Indicators (KPIs). Proposed changes are usually identified to either improve the competitive performance of a process or to increase the competitive attractiveness of an organization's products or services to its targeted market. Strategic objectives in both dimensions are essential in order to have a globally competitive organization. Hoshin Planning is intended to help organizations

- Identify specific opportunities for improvement
- Identify critical business assumptions and areas of vulnerability
- Set or quickly revise strategic vision
- Involve all leaders in planning to set business objectives in order to address the most imperative issues
- Set performance improvement goals for the organization
- Develop change management strategies for addressing business objectives
- Communicate the vision and objectives to all employees
- Hold participants accountable for achieving their part of the vision
- Align and leverage key departments
- Align and leverage suppliers
- Gain and keep customers
- Gain and keep investors

A critical challenge for organizations is often aligning objectives with employee work systems so that they are unified in their pursuit of strategy attainment. Alignment must include linking cultural practices, strategies, tactics, organizational systems, structure, pay and incentive systems, job design, and measurement systems in order to ensure that all elements are working together. At the beginning of the Hoshin cycle, executive leadership sets the organization's overall vision and the annual strategic objectives and targets. At each level moving downward within the organization, managers and employees contribute to that vision and participate in the definition from the overall vision and their annual targets, to the strategy and detailed action plan they will use to attain these targets. They also define the measures that will be used to demonstrate that they have successfully achieved their targets.

Another key principle of Hoshin Planning is to deploy and track only a few priorities at each level of the organization. Given all of the rapid changes and increasing distractions organizations face, individuals must be able to focus on those things that offer the greatest advantage to the organization. The clearer the priorities, the easier it is for people to focus on what truly delivers the most value at any given time.

There are seven steps for effective Hoshin planning:

1. Identify your critical objectives (five-year vision and one-year action plan).

2. Ensure each objective is Specific, Measurable, Attainable, Realistic, and Time Sensitive (SMART).

3. Evaluate any constraints.

4. Establish performance measures.

5. Develop an implementation plan.

6. Assign ownership of goals.

7. Conduct regular reviews.

NOTE *SMART (Specific, Measurable, Attainable, Realistic, and Time Sensitive) is a mnemonic used to guide organizational leaders toward setting specific objectives over and against more general goals. Ultimately, it serves to ensure that all goals are clear, unambiguous, and without vagaries or platitudes. For a goal to be specific it must tell an organization's workforce exactly what is expected, including why it is necessary, who will be involved, where it will happen, and which elements are most pertinent.*

A key technique when using the Hoshin Method is called Catchball. Catchball is a tool used to manage group dialogue and gather ideas. In Catchball, managers and front-line workers develop the strategies, tasks, and metrics to support the accomplishment of the overall Hoshin. This is where the organization and its management team consider what actions are required to achieve each goal. The output of the Catchball process is a detailed action plan for implementation.

Other Hoshin Planning Techniques used to gather ideas and help set targets include:

- *Stop listing:* This is the act of creating a list of things the organization should no longer do. These may be projects that are not realizing value and should be cancelled, markets to get out of, or products that should be discontinued.

- *Voice of customer analysis:* The voice of the customer (VOC) is an extremely valuable part of an organization's planning process. It allows organizations to find valuable insights into what their customers want or need and use those findings as a marketing advantage.

- *Bowling charts:* These are visual depictions of measures and progress toward organizational goals. They are usually monthly measures that outline goals versus actual metrics with stoplight color-coding. In many cases, goals that have not been completed in previous iterations are carried over into the new planning cycles if they are still deemed necessary.

- *Five-year outlook:* The five-year outlook plan is used to help leaders think beyond the organization's current fiscal year. It serves to remind individuals of the longer-term strategic direction the organization wishes to take and which Process Improvement efforts need to be launched in order to achieve these targets.

The Hoshin Planning approach also aims to ensure that insight and vision are not forgotten as soon as planning activities are completed. Plans, once confirmed, are kept alive, acted on daily, and are not abandoned as soon as they have been completed. Regular and iterative reviews take place to identify any progress, discuss any issues, and initiate any needed corrective action. Hoshin Planning implements daily, weekly, monthly, and yearly points of review. Accountability lies with each employee to ensure their goals and targets remain aligned and up to date with the level above them. Sound Hoshin Planning involves continuous communication. Organizations often incorporate feedback systems that allow bottom-up, top-down, horizontal, and multidirectional communication of suggestions, issues, concerns, and impediments.

In closing, Hoshin Planning provides an opportunity to continually improve performance by disseminating and deploying the vision, direction, and plans of corporate management to all employees so that all job levels can continually act, evaluate, study, and provide feedback on the organization's future and its processes. Figure 6-3 is an example of a Hoshin Kanri.

Process Management

An organization with a logical structure will be ineffective if it is not managed appropriately. *Process Management* is the ensemble of planning, engineering, improving, and monitoring an organization's processes in order to sustain organizational performance. It is a systematic approach to making an organization's workflow more effective, more efficient, and more capable of adapting to an ever-changing environment. A business process can be seen as a value chain or a set of activities that will be used to accomplish a specific organizational goal in order to meet customer requirements. Consequently, the goal of process management is to reduce human error and miscommunication in these processes and to focus stakeholders on improving their operating environment using Process Improvement methods. Process Management and Improvement efforts may be initiated from a business or operational unit or the IT organization. These efforts can focus on innovation, transformation, organizational development, change management, enterprise architecture, performance, compliance, and so on.

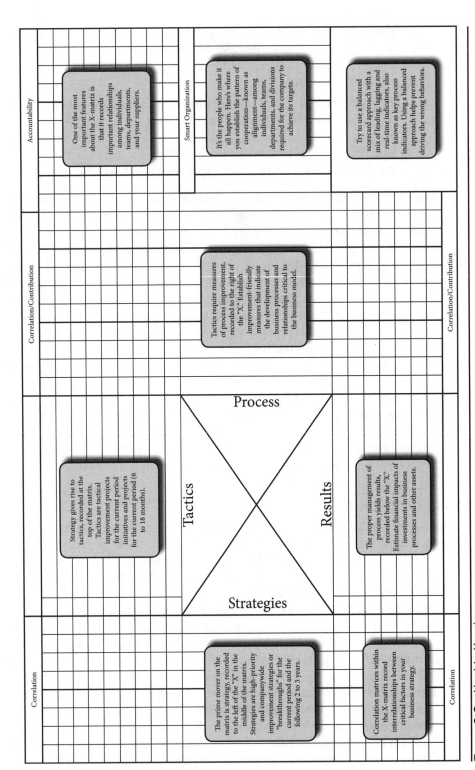

The prime mover on the matrix is strategy, recorded to the left of the "X" in the middle of the matrix. Strategies are high-priority and companywide improvement strategies or "breakthroughs" for the current period and the following 2 to 3 years.

Strategy gives rise to tactics, recorded at the top of the matrix. Tactics are tactical improvement projects for the current period initiatives and projects for the current period (6 to 18 months).

One of the most important features about the X-matrix is that it records important relationships among individuals, teams, departments, and your suppliers.

It's the people who make it all happen. Here's where you establish the pattern of cooperation—known as alignment—among individuals, teams, departments, and divisions required for the company to achieve its targets.

Try to use a balanced scorecard approach with a mix of leading, lagging and real-time indicators, also known as key process indicators. Using a balanced approach helps prevent driving the wrong behaviors.

Tactics require measures of process improvement, recorded to the right of the "X." Establish improvement-friendly measures that indicate the development of business processes and relationships critical to the business model.

The proper management of process yields results, recorded below the "X." Estimate financial impacts of investments in business processes and other assets.

Correlation matrices within the X-matrix record interrelationships between critical factors in your business strategy.

Process

Tactics

Results

Strategies

Accountability

Smart Organization

Correlation/Contribution

Correlation/Contribution

Correlation

Correlation

FIGURE 6-3 Hoshin Kanri

There are several kinds of Process Improvement methodologies being practiced in the market today. Some of the more common methods or management paradigms include Just Do It, Kaizen, Lean Six Sigma, and Rummler–Brache, and are discussed below.

Just Do It

Just Do It is the most basic concept of Process Improvement. This model is used primarily when a problem with a process has been identified, the solution is known and understood, and very little effort is required to implement the change. Often, process issues that fall into this category can be completed within hours or by holding a meeting with key stakeholders to resolve a particular issue. Although Just Do It is a convenient and straightforward method for improving a process, ongoing process monitoring is still required.

Kaizen

Kaizen is a methodology that was established shortly after World War II by the infusion of US quality principles and Japanese philosophies and concepts. *Kai*, meaning to take a part and make new, and *Zen*, to think about in order to help others, make up the philosophy and approach that is the basis for all Process Improvement initiatives.

Kaizen Events are intensely concentrated, team-oriented efforts that are initiated to rapidly improve the performance of a process. A Kaizen Event involves holding small workshops attended by the owners and operators of a process in order to make improvements to the process, which are within the scope of the process participants. Effort is coordinated over a short period of time, usually less than five days, and involves deliverables and activities that must be completed prior to and immediately following the event in order to ensure successful execution. Any solution or task that requires action during or after the Kaizen Event should be easy to implement and not cause significant disruption to the organization. This will demonstrate immediate success and help generate momentum for ongoing improvement efforts.

Purpose of Kaizen Events

The purpose of a Kaizen event is to gather operators, managers, and owners of a process in one place in order to

- Map the existing process
- Use qualitative analysis techniques to determine problems
- Rapidly improve the existing process
- Solicit buy-in from all parties related to the process
- Implement solutions and hold further events for Continuous Improvement

Plan-Do-Check-Act

Among the most widely used tools for Kaizen events is a model known as the *Plan-Do-Check-Act (PDCA) cycle,* also known as the Plan-Do-Study-Act or Deming Improvement Model. The PDCA cycle is a simple, yet powerful, tool for accelerating improvement. The ability to develop, test, and implement change is essential for any individual, group, or organization that wants to continuously improve. The PDCA model is not meant to replace change models that organizations may already be using. Rather, PDCA models accelerate improvement. This model has been used successfully by organizations to improve many different processes and outcomes.

The PDCA model has the following two parts (Figure 6-4):

- Three fundamental questions, which can be addressed in any order:
  - What are we trying to accomplish?
  - What changes can we make that will result in improvement?
  - How will we know that a change is an improvement?
- The PDCA four-step cycle to test and implement changes. This cycle guides improvements through testing in order to determine if the change is appropriate through to implementation.

The PDCA cycle involves the following four steps:

1. Plan: Plan the Kaizen event.
   - Determine the objective and gather information regarding resources, management, and customer complaints about the process in question
   - Gather any current performance data that are available
   - Make predictions about what will happen and why
   - *Estimated time frame to complete: 1–3 weeks*

2. Do: Facilitate the Kaizen Event.
   - Document and analyze the current process
   - Document problems and unexpected observations
   - Determine what modifications should be made
   - Implement improvements to meet management and customer objectives
   - *Estimated time frame to complete: 2–5 days*

3. Check: Set aside time to analyze the data and study the results.
   - Complete analysis of the data related to improvements and overall process performance
   - Compare the data to predictions

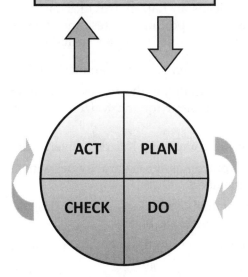

What are we trying to accomplish?	**Team Composition** Including the right people on a process improvement team is critical to a successful improvement effort. Teams vary in size and composition, and are built to suit the improvement effort.
How will we know that a change is an improvement?	**Setting Goals** Improvement requires setting goals. The goals should be time specific, measurable, and identify the population that will be affected
What changes can we make that will result in improvement?	**Selecting Changes** All improvement requires making changes, but not all changes result in improvement. Organizations must identify the changes that are most likely to result in improvement.
	Measurement A critical part of testing and implementing changes. It indicates whether the changes will actually lead to improvement.
ACT PLAN **CHECK DO**	**Testing Changes** Testing a change in a real work setting is always appropriate. Plan it, try it, observe the results, and act on what is learned.
	Implementing Changes After testing a change on a small scale, learning from each test, and refining the change through several PDCA cycles, the change can be implemented on a broader scale.

FIGURE 6-4   PDCA Cycle

- Summarize and reflect on what was learned
- Present results to management and the organization
- *Estimated time frame to complete: 1–2 weeks*

4. Act: Document and standardize new processes and create a monitoring plan to ensure improvements are sustained

- Prepare a plan for another Kaizen Event if necessary
- Implement improvements on a wider scale, if appropriate
- *Estimated time frame to complete: 1–2 weeks*

NOTE *Although Kaizen is specifically outlined as a technique or method for Process Improvement in this chapter, improvement is not the sole purpose of a Kaizen event. Ultimately, Kaizen is meant to teach people how to solve problems (process or otherwise) and engage members of the workforce in doing so. Self-organization around problem solving and infusing a culture where staff members regularly confront and solve problems as a primary part of their job is at the root of the Kaizen approach.*

Roles and Responsibilities for Kaizen Events

This section lists the roles and responsibilities involved in a Kaizen Event.

Process Improvement Manager: The Process Improvement Manager is the primary contact for all improvement efforts surrounding a process that has been selected for improvement using the Kaizen method. The Process Improvement Manager leads all Kaizen activities, facilitates the Kaizen Event, and is accountable for reporting progress and coordinating communication to stakeholders. Other responsibilities include

- Training team members in Kaizen principles and techniques
- Working with management to define the process area, resources, and problem and goal statement for the Kaizen improvement effort
- Scheduling all meetings for completing Kaizen deliverables
- Clearly defining desired outcomes of Kaizen activities with leadership and project team members
- Managing the implementation of solutions and ensuring the transition of the improved process to the business
- Maintaining all documentation from the event and preparing and submitting all deliverables

Kaizen Team Members: Kaizen Team Members are the primary resources assigned to the initiative and are responsible for completing the tasks listed in the action plan, both during and immediately following the Kaizen Event. They are usually operators from the various departments within a process and are considered spokespersons for the Kaizen methodology. They participate in all Kaizen activities as representatives of their operational units. As a rule of thumb, most Kaizen teams do not exceed six members. Other responsibilities include

- Providing process expertise and feedback during all Kaizen activities
- Delivering regular updates to the team and management on the status of deliverables

- Helping to manage the implementation of solutions and ensure the transition of the improved process to the business owner and various operational teams
- Acting as a change catalyst

Subject Matter Experts: A *Subject Matter Expert* (*SME*), often called domain expert, is an expert in a particular process, area, or topic. They have significant knowledge or skills in a particular area or endeavor relevant to the process in question. They participate in Kaizen Events as needed and are also considered spokespersons for the Kaizen methodology. Other responsibilities include

- Providing process expertise and feedback during all Kaizen activities
- Helping to manage the implementation of solutions and ensure the transition of the improved process to the business owner and various operational teams
- Acting as a change catalyst

Management Team: The Management Team is comprised of department managers, executives, and the owner of the process being improved. This team works with the Process Improvement Manager (Kaizen Lead) to identify the areas that require improvement and to determine the objectives of the Kaizen activities. Their primary function is to direct the team toward a goal and approve any deliverables. Other responsibilities include

- Driving Kaizen and the Continuous Improvement culture
- Attending all Kaizen Events as needed and providing approval/ feedback to the team
- Identifying resources and providing time and materials to execute activities
- Publicly endorsing Kaizen improvement activities
- Removing barriers to any Kaizen activities

Guidelines for Kaizen Events
Kaizen events should

- Focus on quality or efficiency problems
- Focus on processes with no more than two outputs
- Focus on processes with no more than two customers
- Require no external resources (e.g., outside customers)

- Fall within the immediate span of control of the Kaizen team
- Be narrowly scoped to include no more than seven to eight stake-holders

Key Kaizen Deliverables

Figure 6-5 outlines several common deliverables for the Kaizen process:

Lean Six Sigma

Lean Six Sigma, which was started in the early to mid 1980s, is a problem-solving and continuous process improvement method that is used to reduce variation in manufacturing, service, or other business processes. Sigma is a Greek letter as well as a statistical unit of measurement used to define the standard deviation of a population. In industry terms, Sigma is a name given to indicate how much of a process's output falls within customers' requirements. The sigma scale ranges from 1 to 6, where the higher the sigma value, the fewer defects that are occurring, the faster the cycle time, and the more effective process outputs are at meeting customer requirements.

Lean Six Sigma projects measure the cost benefit of improving processes that are producing substandard service to customers and primarily focuses on reducing variation. The goal of each successful Lean Six Sigma project is to produce statistically significant improvements in a process.

PLAN	DO	CHECK	ACT
Kaizen Charter	SIPOC	Quality & Process Indicators	Process Procedures
Management Commitment	As-Is Process	Data Collection	Monitoring Plan
Kaizen Team Commitment	Lean Process Analysis	Data Display	Response Plan
Pre-Kaizen Event Kick-Off	Root Cause Analysis	New Process Performance	Replication Opportunities
Kaizen Event Scheduled	Generate Solutions	Dashboard	Solution Transfer Plan
Data Collection	Quick Win Identification	**Management Approval**	**Management Approval**
Data Display	Solution Selection		
Kaizen Event Agenda	To-Be-Process		
Management Approval	Implementation Plan		
	Communication Plan		
	Management Approval		

FIGURE 6-5   Key Kaizen Deliverables

Over time, multiple Lean Six Sigma projects produce virtually defect-free performance and produce substantial financial benefit to an operating business or company.

Lean Six Sigma also focuses on eliminating the seven kinds of wastes, or Muda (pronounced "moo-duh"; a Japanese term that means uselessness, idleness, or wastefulness). A commonly use mnemonic for identifying various wastes within a process is TIMWOOD. The seven waste categories are

- Transportation
- Inventory
- Motion
- Waiting
- Overproduction
- Overprocessing
- Defects

Lean Six Sigma is an amalgamation of Lean's waste elimination characteristics and Six Sigma's focus on the critical to quality characteristics of a process. Formal training does exist for Lean Six Sigma and is provided through the belt-based training system that endorses various levels of expertise and experience from novice to practitioner: White, Yellow, Green, Black, and Master Black. An excellent organization that certifies individuals in Six Sigma practices is the Acuity Institute (www.acuityinstitute.com).

Purpose of Lean Six Sigma

The purpose of a Lean Six Sigma project is to gather operators, managers, and owners of a process in order to

- Ensure that the voice of the customer (Customer Requirement) is understood
- Identify root causes of a problem and initiate improvements
- Identify non–value-added activities in processes
- Standardize existing business processes
- Identify any new process needs
- Coordinate metrics among teams, suppliers, and customers
- Improve customer satisfaction
- Ensure corporate, industry, and government compliance

Define-Measure-Analyze-Improve-Control and Define-Measure-Analyze-Design-Verify

Lean Six Sigma projects follow two models inspired by Deming's Plan-Do-Check-Act Cycle. These models are composed of five phases and bear the acronyms DMAIC (Define-Measure-Analyze-Improve-Control; pronounced "duh-may-ick") and DMADV (Define-Measure-Analyze-Design-Verify; pronounced "duh-mad-vee").

The DMAIC model is used primarily to improve existing processes that are not meeting customer requirements or business objectives. Because these processes often contain several non–value-added activities, a DMAIC project is initiated to identify the root causes of the problem and remove waste from the process. DMAIC consists of the following five phases (Figure 6-6):

- *Define:* Identify customer requirements (voice of the customer) as well as the project goals
- *Measure:* Identify and measure performance of the current process and collect any necessary data
- *Analyze:* Analyze the data to investigate and verify cause-and-effect relationships, determine what the relationships are, and seek out the root cause of the defect under investigation
- *Improve:* Optimize the current process based upon data analysis and design a new, future state process; execute pilot runs to assess process capability
- *Control:* Monitor the newly deployed process to ensure that any deviations from the target are corrected before they result in service issues, implement control systems such as statistical dashboards, and continuously monitor the process

The DMADV model, also known as DFLSS (Design-For-Lean-Six-Sigma), is used primarily to create new processes or improve processes that have been significantly degraded and require enormous effort to improve. A DMADV project is initiated when a new process is needed or when an

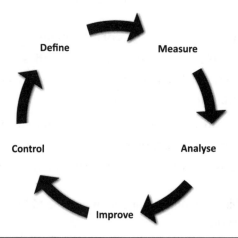

Figure 6-6 The DMAIC Cycle

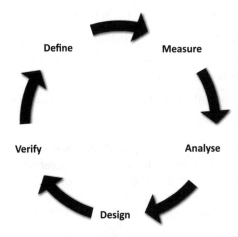

FIGURE 6-7 The DMADV Cycle

existing process requires significant architectural changes. DMADV consists of the following five phases (Figure 6-7):

- *Define:* Design goals that align with customer demands and the enterprise strategy
- *Measure:* Measure and identify CTQ (characteristics that are Critical to Quality), process capability, and risks
- *Analyze:* Develop alternatives in order to create high-level designs and evaluate design capabilities to select the best design
- *Design:* Optimize the design and plan for its verification; this may require the use of simulations
- *Verify:* Verify the design, set up any necessary pilots, implement the process in production, and transition it to the process owner(s)

Roles and Responsibilities for Lean Six Sigma Projects

Six Sigma roles are structured to support project success and drive business performance results. It is critical to gain agreement regarding roles and time commitment early in the process so that teams can establish chemistry and focus on project deliverables.

Executive Leadership: Executive Leadership provides and aligns resources for any improvement initiative, specifically at the sponsor level. The executive leadership team defines the business strategy and establishes improvement priorities and targets and is often organized within a formal steering committee. Other responsibilities include

- Monitoring internal and external factors that are affecting the business
- Communicating the plan for overall business success

- Championing the Lean Six Sigma vision
- Establishing accountability for results
- Setting the tone, serving as role models who display appropriate behaviors, and acting as change leaders
- Integrating Lean Six Sigma into business direction and plan
- Marketing Lean Six Sigma programs and results

Process Owner: The Process Owner's role exists to provide expertise on a process and to provide resources to serve as team members and SMEs on projects. They are the primary proprietor of any items related to the process in question and assist with determining Lean Six Sigma projects and goals. Other responsibilities include

- Approving and supporting Six Sigma projects
- Approving changes in project scope and removing barriers
- Owning the solution being delivered by the project team
- Supporting the implementation of improvement actions

Project Sponsor: The Project Sponsor's role exists to provide and align resources, ensure that project deliverables are being maintained, and ensure that the project is delivered on time and on budget. They also ensure that the project is aligned with department and strategic objectives and establish improvement priorities, targets, and accountability results. Other responsibilities include

- Ensuring cross-functional collaboration
- Role modeling appropriate behaviors
- Acting as a change leader
- Approving all phases of a Lean Six Sigma project
- Approving changes in project scope and removing barriers
- Marketing Lean Six Sigma programs and results

NOTE *In many cases, the Process Owner and Project Sponsor are the same person supporting a Six Sigma Project.*

Process Architect: The Process Architect provides expertise on Lean Six Sigma tools and techniques as well as process architecture and change management principles. They are responsible for ensuring that any changes made within a particular process conform to Process-Oriented Architecture (POA) principles and do not disrupt the Process Ecosystem in a negative manner. This position can vary from full-time to part-time assignment depending on an initiative's complexity. These individuals train and coach process improvement managers and project team members and guide them on how to properly design processes that meet the enterprise's strategic objectives. They are considered advisors to project champions, sponsors, and process owners. Other responsibilities include

- Providing strategic direction to leadership and project teams
- Identifying projects that are critical to achieving business goals and sustaining an organization's process ecosystem
- Serving as the main champion during change implementation
- Ensuring cross-functional and cross-team collaboration
- Ensuring that all appropriate process architecture principles and standards are followed

Process Improvement Manager: The Process Improvement Manager provides direction and leadership for project teams, directs interproject communications, and manages the day-to-day activities of a Lean Six Sigma Process Improvement project. This is a full-time position that is accountable for reporting project progress and coordinating communication to stakeholders. These individuals directly manage the implementation of solutions and ensure the transition of improved processes to the business. Other responsibilities include

- Maintaining all documentation for a project and preparing and submitting deliverables
- Delivering results through the application of the Lean Six Sigma methodology
- Providing skills training when needed
- Acting as a change catalyst

Project Team Members: Team members are the primary resources assigned to the initiative and are responsible for tasks within the project's action plan. They are considered spokespersons for the Lean Six Sigma methodology and participate in project activities as needed. Other responsibilities include

- Providing process expertise and feedback during all Process Improvement activities
- Delivering regular updates to the team and management on the status of action tasks
- Helping to manage the implementation of solutions and ensuring the transition of the improved process to business owners and operational teams
- Acting as a change catalyst

Subject Matter Experts: As with Kaizen Events, SMEs are assigned to Lean Six Sigma Process Improvement projects and are considered domain experts or persons with significant knowledge or skills in a particular area of endeavor relevant to the process in question. They participate as needed in Lean Six Sigma projects and are crucial for understanding the true dynamics within the operational units involved with a process. Other responsibilities include

- Providing process expertise and feedback during any necessary project activities
- Helping to manage the implementation of solutions and ensuring the transition of the improved process to business owners and operational teams
- Acting as a change catalyst

Financial Analyst: The Financial Analyst is responsible for providing financial support to Lean Six Sigma projects. They are involved on an as-needed basis and have the primary function of providing financial savings forecasts for deliverables and changes within a process or the project as a whole. Other responsibilities include

- Providing standard and consistent guidelines for project valuation
- Estimating project savings during project execution
- Tracking and validating actual project savings after project closure

Guidelines for Lean Six Sigma Projects

Lean Six Sigma projects should

- Focus on business value
- Be linked to the enterprise strategy and support specific business-centric objectives
- Be viable, visible, and verifiable
- Contain Goals and a vision for the improvement effort that is endorsed by leadership
- Conduct frequent and iterative reviews of goals and timelines
- Ensure that proper resource planning (top-down or bottom-up) occurs
- Hold leads and participants accountable for their deliverables
- Utilize proper governance and decision criteria for decision-making

Key Lean Six Sigma Deliverables

Figure 6-8 outlines several common deliverables for Lean Six Sigma Projects:

DEFINE	MEASURE	ANALYSE	IMPROVE	CONTROL
☐ Project Charter	☐ Identify Measures	☐ Root Cause Analysis	☐ Identify & Select Solutions	☐ Process Control
☐ Project Management	☐ X/Y Matrix	☐ Affinity Diagram	☐ Generate Solutions	☐ Control Charts
☐ Project Plan	☐ Data Collection	☐ Fishbone Diagram and/or 5 Why's	☐ Benchmarking	☐ Process Monitoring Plan
☐ Process Definition	☐ Operational Definitions	☐ Lean Process Analysis	☐ Solution Prioritization Matrix	☐ Dashboard
☐ Process Profile/ SIPOC	☐ Sample Size Calculations	☐ Lean Tools & Measures	☐ Solution Selection Matrix	☐ Response Plan
☐ As-Is Process Map	☐ Measurement Systems Analysis (MSA)	☐ Graphical Data Analysis	☐ To-Be Process map	☐ Project Documentation
☐ Quick Win Identification	☐ Data Collection Plan	☐ Histogram	☐ Financial Impact of Solutions	☐ Process Procedures
☐ Stakeholder Management	☐ Describe & DisplayData	☐ Pareto Chart	☐ Cost/Benefit Analysis	☐ Replication Opportunities
☐ Stakeholder List	☐ Histogram	☐ Box Plot	☐ Risk Planning & Testing	☐ Solution Transfer Plan
☐ Stakeholder Management Plan	☐ Pareto Chart	☐ Multi-Var	☐ Failure Modes & Effects Analysis (FMEA)	
☐ Communication Plan	☐ Pie Chart	☐ Correlation Analysis	☐ Pilot Plan	
☐ Voice of Customer	☐ Run Chart	☐ Statistical Data Analysis	☐ Statistical Data Analysis	
☐ Customer Identification	☐ Control Charts	☐ Linear Regression	☐ Implement Solutions	
☐ VOC Research Plan	☐ Baseline Performance	☐ Multiple Regression	☐ Multi-Generational Project Plan	
☐ Kano Analysis	☐ Sigma performance	☐ t-Test	☐ Implementation Plan	
☐ CTQ Identification	☐ Yield	☐ ANOVA	☐ Stakeholder Management	
	☐ Process Capability	☐ Chi-Square	☐ Project Storyboard	
	☐ Quick Win Identification	☐ Design of Experiments		
		☐ Root Causes Identified		
		☐ Quick Win Identifications		

FIGURE 6-8   Key Lean Six Sigma Deliverables

135

Rummler–Brache

Geary Rummler and Alan Brache defined a comprehensive approach to organizing companies around processes, managing and measuring performance, and redefining processes in their book, *Improving Performance: How to Manage the White Space on the Organization Chart*. The *Rummler–Brache Methodology* is a systematic approach to business process change. It is a step-by-step set of instructions on how to make changes to the way in which work is completed across an organization. The primary differentiator of the Rummler–Brache model is the premise that all organizations behave as adaptive systems with three levels of focus (the organization, its processes, and its workforce), whereby any change being proposed in an improvement effort must take into account all three variables during process analysis and design.

Purpose of Rummler–Brache

The purpose of a Rummler–Brache Project is to gather operators, managers, and owners of a process in order to

- Address performance in a comprehensive (rather than piecemeal) fashion
- Identify disconnects within a process (people, process, technology) and their root causes in order to initiate improvements
- Ensure the supplier, customer, employee, and shareholder experience is understood
- Identify new process needs
- Improve customer satisfaction and ensure that measurable results exist for customers and stakeholders
- Identify and improve how the workforce interacts and contributes to processes and ensure that the human component is at the forefront of any proposed change (user-centric design)

The Rummler–Brache methodology is based on the following two core concepts:

- *Three Levels of Performance:* These levels are the organization, process, and performer, or people, levels. To effect change in an organization, it is necessary to understand the potential impacts to all three levels. For example, a process change could mean significant changes to job responsibilities and skills required to execute those responsibilities. Failure to adequately account for these interrelationships is a leading cause of failed process improvement implementations.

- *Three Performance Dimensions:* These dimensions are goals, design, and management. Having clear goals at each level ensures alignment with desired results; having robust design at each level maximizes the efficiency of operations; and having good management systems at each level ensures that the organization can survive and adapt to changes in the business environment. A failure in any of these dimensions will lead to performance problems.

Rummler–Brache Model

Rummler–Brache projects follow a model inspired by Deming's Plan-Do-Check-Act Cycle. This model is composed of the following five phases and bears the acronym ADDDSA (Assess-Define-Develop-Deploy-Sustain-Adapt; pronounced "Ad-sah"):

- *Phase 0, Assess:* The Assess phase seeks to understand the customer's voice and the customer's true requirements for the organization. It also examines the organization's processes to determine which processes require intervention and which processes simply need to be maintained based on these requirements.

- *Phase 1, Define:* The Define phase determines the boundaries of the process or system of processes that will be worked on. It sets performance targets and lays out a road map for achieving these targets. This information, plus information about stakeholders and change management strategies, is captured in the Define phase.

- *Phase 2, Develop:* The Develop phase designs and tests modifications to processes. These modifications can range from basic streamlining to a green field approach where new processes are created. The two main requirements are the analysis and mapping of the existing process (as-is) and the design of the future-state process (should). Disconnects in the current-state or as-is process are clearly listed, prioritized, and defined prior to future-state solution design.

- *Phase 3, Deploy:* The Deploy phase manages the adoption, stabilization, and institutionalization of a new process and/or any changes needed to rectify disconnects found in the Develop phase. This can be done in iterations to maximize adaptability and minimize risk. Key activities of the Deploy phase include executing the implementation plan as well as focusing on communication and change management. Special attention is put on Human performance and the effects of change on the organization's workforce.

- *Phase 4, Sustain:* The Sustain phase confirms that what has been implemented is achieving the expected results and continues to improve any remaining disconnects. The activity in this phase primarily involves process monitoring.

- *Phase 5, Adapt:* The Adapt phase transitions the process to the necessary business owner or operations team and trains owners and operators on how to monitor the process on an ongoing basis. It ensures that metrics are created and in place and that they align with the organization's strategic goals.

Roles and Responsibilities for Rummler–Brache Projects

The Rummler–Brache life cycle is similar to that of most Process Improvement methods and, as such, its project makeup is comprised of the following key roles:

Executive Leadership: Executive Leadership provides and aligns resources for any improvement initiative, specifically at the sponsor level. The executive leadership team defines the business strategy and establishes performance goals and improvement priorities.

Project Sponsor: The Project Sponsor's role exists to ensure project deliverables are being maintained and that the project is delivered on time and on budget. They also ensure that the project is aligned with department and strategic objectives as well as any improvement priorities. The project sponsor is ultimately accountable for a project's results.

Process Owner: The Process Owner's role exists to provide expertise on a process and to provide resources to serve as team members and SMEs on projects. They are the primary proprietor of any items related to the process in question and assist with determining Rummler–Brache projects and determining the goals for them.

Process Architect: The Process Architect provides expertise on the Rummler–Brache method as well as process architecture and change management principles. They are responsible for ensuring any changes made within a particular process conform to POA principles and do not disrupt the organization's Process Ecosystem in a negative manner. These individuals train and coach process improvement managers and project team members and are considered advisors to project champions, sponsors, and process owners.

Process Improvement Manager: The Process Improvement Manager leads the process owner and executive team in defining the scope of the process improvement and its goals, boundaries, staffing, and timetable. The role demands a high degree of planning and organizing skills and the ability to advise and coach senior members.

Project Team Members: Project Team Members are the primary resources assigned to the initiative and are responsible for tasks within the project schedule. They are considered spokespersons for the Rummler–Brache methodology and participate in project activities as needed.

Guidelines for Rummler–Brache Projects

Rummler–Brache projects should

- Focus on improving overall performance
- Ensure that all disconnects within a process, including people and technology gaps, are identified and considered
- Be linked to the enterprise's strategic objectives
- Be endorsed by leadership
- Conduct frequent and iterative reviews of goals and timelines
- Ensure that proper resource planning (top-down or bottom-up) occurs
- Hold participants accountable for their deliverables
- Utilize proper governance and decision criteria for decision-making

Key Rummler–Brache Deliverables

Figure 6-9 outlines several common deliverables for Rummler-Brache projects:

Selecting a Process Improvement Methodology

Professionals have often debated or questioned when to initiate a full-scale Six Sigma or Rummler–Brache project and when to initiate smaller-scale improvement efforts such as Kaizen Events. Although the models are similar, there are distinct reasons for initiating one methodology over the other. Kaizen is usually used for small-scale or urgent Process Improvement initiatives where the root cause of a problem is widely understood, but the solution is unknown. A two- to five-day Kaizen workshop is initiated to analyze this process and to plan and implement improvements to it. Lean Six Sigma is used to create new processes or to analyze existing processes where the problem is not yet understood. Lean Six Sigma determines the root cause of a problem and uses statistical analysis to determine the most optimal way to improve the issue. Rummler–Brache is also used to create new processes or to analyze existing processes where problems are not yet understood. Where Lean Six Sigma places heavy emphasis on statistical analysis and measurements, Rummler–Brache focuses more on workforce or human performance indicators. Ultimately, the methodology that is chosen depends on the complexity of the business problem, what the organization's focus is (workforce or metric), and what method best suits the stakeholders being serviced. In many cases, practitioners often combine several techniques from multiple methods in their performance planning and improvement efforts. Figure 6-10 identifies the key decisions involved when selecting a methodology for improvement.

Phase 0: Access	Phase 1: Define	Phase 2: Develop	Phase 3: Deploy	Phase 4: Sustain	Phase 5: Adapt
☐ Strategy Document	☐ Project Charter	☐ As-Is Process Map	☐ Implement Solutions	☐ Performance Metrics List	☐ Training Guides
☐ Super System Map	☐ Steering Committee Team	☐ Disconnects Identified	☐ Stakeholder Management	☐ Process Monitoring Plan	☐ Transition Plan
☐ Voice of Customer	☐ Process Improvement Team	☐ Root Cause Analysis	☐ Implementation Team Structure	☐ Dashboard	☐ Retrospective
☐ Critical Business Issues (CBI) List	☐ Performance Goals & Metrics	☐ Should Process Map	☐ Human Performance & Workforce Analysis	☐ Issue Response Plan	☐ Project Summary Report
☐ Critical Success Factors (CSF) List	☐ Critical Process Issue (CPI) Identification List	☐ Quick Win Identification	☐ Start/Stop/Keep	☐ Process Procedures	
☐ Process Relationship Map	☐ Process Boundaries	☐ Recommended Change Log	☐ Pilot	☐ Replication Opportunities	
☐ Process Priority List	☐ Process Profile	☐ Solution Selection	☐ Change Management		
☐ Action Plan	☐ Stakeholder Map	☐ Implementation Plan			
	☐ Communication Plan				
	☐ Assumptions & Constraints				

Figure 6-9 Key Rummler–Brache Deliverables

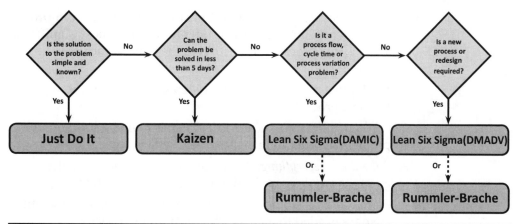

FIGURE **6-10** Selecting a Process Improvement Methodology

Other Process Improvement Methodologies

Most Process Improvement methods contain several valuable philosophies and approaches that can be used in any Process Improvement project. In many cases, certain philosophies can be combined with other methods to achieve greater process excellence. Beyond the more common methods outlined in detail in this chapter, there are several other methodologies that exist in the industry today, including

- *Total Quality Management.* Total Quality Management (TQM) is a Process Improvement approach with a primary focus on customer satisfaction. Conceived from the various works of W. Edwards Deming, Kaoru Ishikawa, and Joseph M. Juran, TQM involves all members of an organization when improving its processes, products, services, and culture. It infuses the following basic principles into the culture and activities of an organization:

  - *Customer focus:* The customer determines the definition of quality for an organization.

  - *Employee involvement:* All employees participate in Process Improvement activities and work toward common goals.

  - *Process centered:* The organizations focuses on process thinking.

  - *Integrated system:* An organization's departments are viewed as interconnected.

  - *Strategic approach:* The organization strategically plans its vision, mission, and goals.

  - *Fact-based decision-making:* Process performance data are viewed as necessary.

  - *Communication:* Effective communication is critical to maintaining morale and motivating employees.

- *Theory of Constraints.* Theory of Constraints is a Process Improvement approach that focuses on identifying the most significant bottleneck in a process and rectifying the issue. It includes the following five-step process:

  1. *Identify the constraint.* The first step is to identify the weakest link in a process.

  2. *Exploit the constraint.* The second step is to find a solution to the problem in order to increase process efficiency.

  3. *Subordinate everything else to the constraint.* The third step is to align the whole organization to support the solution.

  4. *Elevate the constraint.* The fourth step is to make any other changes needed to increase the constraint's capacity.

  5. *Go back to step 1.* The final step is to review how the process is performing with the fix that has been implemented. If the issue has been resolved, the process can be repeated.

- *Streamlined Process Improvement (SPI).* SPI, sometimes called process redesign, focused improvement, process innovation, or process streamlining, is a step-by-step method for improving business processes created by H. James Harrington. It is a systematic way of using cross-functional teams to analyze and improve the way a process operates by improving its effectiveness, efficiency, and adaptability. SPI's primary focus is on upgrading major organizational processes using a five-step approach known as PASIC. Where traditional methods focus primarily on continuous process improvement, SPI believes in a combined organizational strategy to Process improvement that includes both traditional continuous improvement tactics and radical redesign efforts. When breakthrough improvement and continuous improvement are combined, the result is often significant improvement in process performance over continuous improvement alone. Other benefits of Streamlined Process Improvement include

  o Lowering organizational costs

  o Moving decision-making closer to the customer

  o Shortening return on investment cycle

  o Increasing market share

  o Improving productivity

  o Minimizing the number of contacts that customers have

  o Increasing profits

  o Improving employee morale

  o Minimizing bureaucracy

o Reducing inventory

o Improving quality

There are five phases of SPI, which are known as PASIC, and they consist of 31 activities.

o Phase I: Planning for Improvement

- Activity 1: Define Critical Business Processes
- Activity 2: Select Process Owners
- Activity 3: Define Preliminary Boundaries
- Activity 4: Form and Train the PIT
- Activity 5: Box in the Process
- Activity 6: Establish Measurements and Goals
- Activity 7: Develop Project and Change Management Plans
- Activity 8: Conduct Phase I Tollgate

o Phase II: Analyzing the Process

- Activity 1: Flowchart the Process
- Activity 2: Conduct a Benchmark Study
- Activity 3: Conduct a Process Walk-Through
- Activity 4: Perform a Process Cost, Cycle Time, and Output Analysis
- Activity 5: Prepare the Simulation Model
- Activity 6: Implement Quick Fixes
- Activity 7: Develop a Current Culture Model
- Activity 8: Conduct Phase II Tollgate

o Phase III: Streamlining the Process. Refining the process

- Activity 1: Apply Streamlining Approaches
- Activity 2: Conduct a Benchmarking Study
- Activity 3: Prepare an Improvement, Cost, and Risk Analysis
- Activity 4. Select a Preferred Process
- Activity 5: Prepare a Preliminary Implementation Plan
- Activity 6: Conduct Phase III Tollgate

o Phase IV: Implementing the New Process. Installing the new process

- Activity 1: Prepare a Final Implementation Plan
- Activity 2: Implement New Process
- Activity 3: Install In-Process Measurement Systems

- Activity 4: Install Feedback Data Systems
- Activity 5: Transfer Project
- Activity 6: Conduct the Phase IV Tollgate
  - Phase V: Continuous Improvement. Making small-step improvements
    - Activity 1: Maintain the Gains
    - Activity 2: Implement Area Activity Analysis
    - Activity 3: Qualify the Process
- *Toyota Production System.* The Toyota Production System, often referred to as the Lean Manufacturing System or Just-in-Time System, is a Process Improvement approach that strives to eliminate overburden (Muri), inconsistency (Mura), and waste (Muda). It abides by several principles that, together, form the Toyota Way of Thinking. These principles are the basis for the company's Process Improvement efforts. They are
  - Base your management decisions on a long-term philosophy, even at the expense of short-term financial goals.
  - Create a continuous process flow to bring problems to the surface.
  - Use pull systems to avoid overproduction.
  - Level out the workload.
  - Build a culture of stopping to fix problems in order to get quality right the first time.
  - Realize that standardized tasks and processes are the foundation for continuous improvement and employee empowerment.
  - Use visual control so that no problems are hidden.
  - Use only reliable, thoroughly tested technology that serves your people and processes.
  - Grow Leaders who thoroughly understand the work, live the philosophy, and teach it to others.
  - Develop exceptional people and teams who follow your company's philosophy.
  - Respect your extended network of partners and suppliers by challenging them and helping them improve.
  - Go and see for yourself in order to thoroughly understand the situation.
  - Make decisions slowly by consensus, thoroughly considering all options, and implement decisions rapidly.
  - Become a learning organization through relentless reflection and continuous improvement.

- *Business Process Reengineering.* Business Process Reengineering (BPR) is a Process Improvement approach that prescribes totally rethinking and redesigning processes in order to achieve improvements. It advocates that organizations go back to the basics and reexamine their beginnings. BPR is not usually seen as a continuous improvement method but rather as a method for organizations that need dramatic and exponential improvement in short order (Reinventing). BPR also focuses on processes rather than on tasks, jobs, or employees. It endeavors to redesign the strategic and value-added processes that exist within an organization.

Agility and Process Improvement

To remain competitive, organizations, regardless of industry segment, must produce products and services that are of consistently high quality and that continually meet the needs of an ever-changing market. In order to do this, organizations must possess the ability to continuously monitor market demand and quickly respond by improving or enhancing processes in order to provide new products, services, and/or information. Agility is a way to assist organizations in this effort and is an iterative method for delivering projects, products, and services faster. The most popular Agile methodologies include Scrum, Extreme Programming, Lean Development, and Rational Unified Process; all have been primarily focused on software development initiatives to date. The primary objective of Agile is to incorporate iteration and continuous feedback in order to drive project efforts and successively deliver solutions. This involves continuous planning, continuous testing, continuous integration, and other forms of continuous evolution of both the project goals and the project's deliverables. Agility also focuses on a series of principles such as empowering people to collaborate and make decisions together quickly and effectively, delivering business value frequently, and promoting self-organizing teams.

Many of the practices promoted by Agile development can also be applied to Process Improvement activities. For example, one of the most significant reasons for the failure of improvement projects is that project teams focus significant attention on completing tasks in a waterfall-like fashion, whereby all deliverables from one phase of a project must be completed before a project can move into the next phase. This practice can significantly slow down improvements and hinder organizational progress. Application of Agile techniques in a Process Improvement project allows project sponsors and improvement teams to regularly prioritize improvement activities and allows teams to focus on individual improvements or group-related improvements into packages for faster resolution.

Upon analyzing Agile, many of the principles from the Agile Manifesto can be repurposed for Process Improvement and can address many

Agile Principles	Process Improvement Principles
Our highest priority is to satisfy the customer through early and continuous delivery of valuable software	Our highest priority is to satisfy the organization through rapid delivery of useful and valuable process improvements
Welcome changing requirements, even late in development. Agile processes harness change for the customer's competitive advantage	Welcome changing priorities within the Process Improvement projects in order to ensure improvement efforts deliver valuable and timely solutions
Deliver working software frequently, from a couple of weeks to a couple of months, with a preference to the shorter timescale.	Deliver Process Improvements frequently by prioritizing disconnects individually or in smaller groupings to deliver in a couple of days to weeks.
Working software is the primary measure of progress.	Process Improvements which enhance performance are the primary measure of progress.
Agile processes promote sustainable development. The sponsors, developers, and users should be able to maintain a constant pace indefinitely.	Process Improvement & the POA promote sustainable process development. The sponsors, Process Improvement Organization, and process operators should be able to maintain a constant pace indefinitely.

FIGURE 6-11 Agility and Process Improvement

key pitfalls of improvement projects. Figure 6-11 displays several key operating principles from the Agile Software development manifesto and several commonalities and comparative uses in Process Improvement activities.

Performance Management

Performance Management is the basis for sound and rigorous Process Improvement efforts. In order for an organization to have good process and improvement controls, it must be able to see where the organization is truly performing. Having a well-defined set of performance metrics provides management with the means to measure the performance of current processes and measure the success of prior improvement efforts. It also allows management to determine what new areas require consideration in order to improve the effectiveness and efficiency of processes within an organization. More concisely, Performance Management is the

☐ Meeting	☐ Partially Meeting	☐ Not Meeting	☐ Not Applicable
Clear evidence that performance meets or exceeds the standard	Clear evidence that performance meets some, but not all of the standard	Clear evidence that performance does not meet the standard	The item is not applicable

Figure 6-12 Performance Rating Scale

ongoing process of ascertaining how well or, how poorly, an organization and its processes are performing. It involves the continuous collection of information on progress made toward achieving objectives as well as the determination of the appropriate level of performance, the reporting of performance information, and the use of that information to assess the actual level of performance against the desired level.

Ongoing Performance Management helps answer the following five key questions:

- What is the current process performance level and is it meeting customer requirements? A sample performance rating scale is outlined in Figure 6-12.

- Has performance changed?

- Do we need to adjust the process?

- Do we need to improve the process?

- How will the process perform in the future?

Key Performance Metrics

A performance metric is a specific measure of an organization's activities. Performance metrics support a range of stakeholder needs, from customers and shareholders to employees. Although many metrics are often financially based, inwardly focusing on the organization's performance is also critical to ensuring customer requirements are being met. Likewise, an organization can measure almost anything within its processes and operations. Some examples include quantity of defective products produced, cycle time to process an order, and number of outstanding invoices.

Performance metrics can be broken down into the following four distinct levels:

- *M0: The Organization Level.* These metrics measure the achievement of strategic goals and monitor the organization's overall return on investment.

- *M1: The Process Level.* These measures ensure processes deliver the required value at the required efficiency to meet strategic goals.

- *M2: The Subprocess Level.* These metrics ensure optimum performance of key value-added process steps and alignment with downstream process steps.

- *M3: The Performer Level.* These measures ensure that the workforce is aligned with the overall process and business goals.

Process Monitoring Plans

Perhaps the most vital tool used for performance management and monitoring is the Process Monitoring Plan. The monitoring plan clarifies how the process performance will be continuously monitored, who will be notified if there is a problem, how that will happen, and what response is required.

The first part of the monitoring plan specifies the metrics that will be tracked to summarize process performance. It also specifies the process steps, where the process will be measured, how the process will be measured (a popular method is to use control charts to display causes of variation), and how often they will be tracked. Another critical step is to clarify who is responsible for conducting this effort; although the responsibility usually falls to the process owner. The monitoring plan also indicates what constitutes satisfactory performance and what should be considered a red flag, indicating possible problems. Figure 6-13 illustrates an Process Monitoring Plan.

Other questions to be answered when creating a comprehensive monitoring plan include

- Why is the organization interested in this variable? What decisions or actions will it inform?
- What activities or events generate potential data points?
- What should be the unit of measurement?
- How frequently should the data be reviewed?
- Are there hidden variables that can have a significant impact on the value of a metric over time?
- Should the data be drilled down to a deeper level?

The measures or indicators selected by an organization should best represent the factors that lead to improved customer, operational, and financial

Critical Process Measure	Process Step Where Process Element is Measured	Data Collection Method	Data Collection Frequency	Owner Responsible for Collection

FIGURE 6-13   Sample Process Monitoring Plan

performance (a Balanced Scorecard approach). A comprehensive set of performance measures that is tied to customer and organizational performance requirements provides a clear basis for aligning all processes with organizational goals. Quite often, measures and indicators are needed to support decision-making in corporate environments.

Measurements should originate from business needs and organizational strategy, and they should provide critical performance data and information about an organization's key processes, outputs, and results. Several types of data and information are required for performance management including product, customer, and process performance; comparisons of market, operational, and competitive performance; supplier, workforce, partner, cost, and financial performance; and governance and compliance outcomes. Regular analysis and structured reviews of performance data should be conducted and may involve changes to better support an organization's goals as time goes on.

NOTE *A Balanced Scorecard (BSC) is a structured dashboard that can be used by managers to keep track of the execution of activities by the staff within their control and to monitor the consequences arising from these actions.*

Process Dashboards

The Process Dashboard is a key communication tool that an organization uses to summarize process performance. It indicates the overall health of a process in a concise manner by providing a visual picture of the process through data diagrams and charts. It is primarily used for executive reviews but also serves as an excellent tool for management and process owners. The Process Dashboard provides the means for detecting defects and problems and for troubleshooting any variation that may exist in a process. The indicators and gauges used on Process Dashboards should tie to process goals, customer requirements, and strategic objectives. Process Dashboards help effectively communicate process performance to operators, champions, and stakeholders. Figure 6-14 outlines a basic Process Performance Dashboard.

In planning for the design of a process dashboard, practitioners should select the charts that best display process performance and results. To do this, practitioners must first decide what the process dashboard should say or describe about the process it is monitoring. Some key questions to consider when developing a Process Dashboard are

- Is my process capable and is it meeting customer requirements?
- If it is not meeting customer requirements, what might be driving the lack of capability?
- How much variation is in my process?

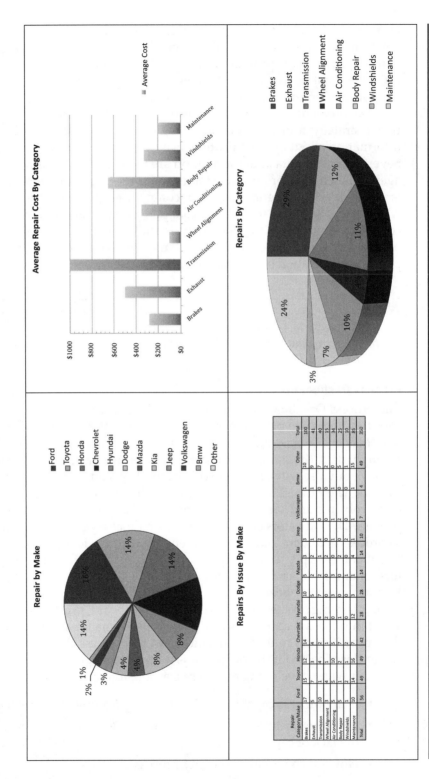

FIGURE 6-14 Sample Process Performance Dashboard

- What types of metrics should I include (e.g., cycle times, sales figures, accuracy rates, sigma levels, defect rates, repetition rates)?
- How do I want the information displayed to my stakeholders and my leadership team (Steering Committee)?
- To what level do I want to drill down in the information?
- How might I want to segment the information for making critical decisions?
- Who should have access to the information?

Once the basic format for the performance dashboard has been decided, it is time to gather the relevant data. Figure 6-15 provides an overview of some of the general charts and tools used for displaying information in a performance dashboard.

Chart/Tool	Type	Purpose
Bar Chart	Display	Used to display simple comparisons between series of data
Cause & Effect Diagrams	Analysis	Used to brainstorm and logically organize possible causes for a particular effect
Control Charts	Monitoring	Used to help understand variation and control and improve a process
Histogram	Monitoring	Used to display frequency of data in column form. Also used to prepare Pareto charts
Pareto Analysis	Monitoring/Analysis	Used to identify and prioritize problems or causes of problems
Pie Chart	Display	Used to depict percentages of total in a circular diagram
Regression	Analysis	Used to track the relationship between two sets of data over time
Run Charts	Monitoring	Type of chart used to display data related to a process variable in sequence over time to identify changes
Scatter Plot	Analysis	Used to confirm relationship between two variables

FIGURE 6-15 Performance Dashboard Charts and Tools

Process Enablers

In order to truly assess and improve process performance, process stakeholders must understand why a process might be performing in a particular manner. Once measures have been established and owners and monitoring plans have been implemented, organizations can then begin to determine the reasons for process performance and initiate subsequent improvements if necessary. The following six common process enablers are used to determine how a process behaves:

- *Process Design:* The activities, sequence, decisions, and handoffs within a process
- *Technology:* The systems, information, applications, data, and networks used within a process
- *Morale and motivation:* The organization's systems that are concerned with how people, departments, and processes are measured and the associated consequences (reward and punishment) that go along with them
- *Workforce:* The knowledge, training, competencies, skills, and experience of the workforce as well as the organizational structure and job definitions of employees
- *Policies and rules:* The Policies, laws, regulations, and associated Business Rules established by the organization to guide or constrain business processes
- *Facilities:* The location, building, furniture, lighting, air quality, and overall workplace design

Workforce Management

Workforce Management describes an organization's system for developing and assessing the engagement of its workforce, with the aim of enabling and encouraging all members to contribute effectively and to the best of their ability. Workforce Management systems are intended to foster high performance, address core competencies, help accomplish action plans, and ensure organizational sustainability.

Human Performance System

Among the most widely used Workforce Management systems is a model known as the Human Performance System, or HPS, designed by Rummler–Brache. The HPS describes the variables that influence a person's behavior in a work system. Performance analysts and other process and performance specialists have used HPS to diagnose and even predict the likely behavior of human beings in given performance situations.

The model is based on three key tenets. These tenets state that every organization is a complex system designed to transform inputs into valued outputs for customers; every performer within an organization and at any level is a part of a unique personal performance system; and when an individual fails to produce a desired outcome, it is due to the failure of one or more components of that person's HPS. In general, HPS is used as a model for understanding the causes of undesired performance and for designing productive, positive work environments that ensure reliable results.

HPS is an excellent tool to use when

- Organizations are not sure why performers are inconsistent in achieving results.
- Different branches, departments, or teams are achieving differing results and there is no clear reason as to why.
- There are disagreements or varying theories about why performers are not achieving reliable results.

In the context of Process Improvement, the job performer is only one of several variables that determine results. Since the job performer usually wants to do well, organizations must carefully examine all factors that could be contributing to a particular performance situation (Figure 6-16). These include

- *Output Expectations:* Do the performers know what is expected? Are the expectations achievable?
- *Inputs/Triggers:* Is it clear when the task or job should be performed? Are there directions? Once the performer starts the task, is there any task interference?
- *Resources:* Does the performer have all the necessary equipment, tools, facilities, and similar items to perform as expected? Are the resources readily available and in ample supply? Are the resources in reasonable condition?

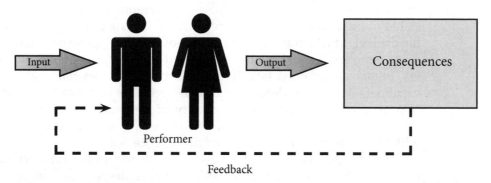

FIGURE 6-16   Human Performance Model

- *Consequences:* When the performer tries to perform as expected, is the result/reaction positive or negative to the performer? What are the immediate consequences versus longer-term consequences?

- *Feedback:* Does the performer get regular feedback on his/her performance? Is the feedback understandable, useful, and relevant? Is the feedback specific to the performer's task/job?

- *Performer:* Has the performer been adequately trained or coached to do this job? Is the performer willing to do the work? Does the performer have the basic capacity (i.e., physical, mental requirements) for this type of work?

Performance deficiencies often appear to be caused by an individual or group of individuals and are actually the result of other disruptions in their HPS, such as

- Missing or insufficient materials
- Unclear direction or expected output
- Interference while trying to do the work
- Absence of performance feedback
- Strong negative consequences for errors
- Lack of positive praise or feedback for succeeding
- Broken, inaccessible, or obsolete equipment
- Lack of training or other preparation
- Lack of motivation or inspiration
- Poor morale
- Lack of purpose or direction

Quality Management

In today's marketplace, quality is known as a measurement of excellence or as a state of being free from deficiencies or significant variations against stated or implied customer requirements. This is determined by measuring the level at which an organization is able to satisfy these requirements. Since performance and quality are judged by an organization's customers, organizations must take into account all processes that may touch customers in order to determine if their needs are being met. In Process Improvement organizations, the quality of the processes used to meet these customer requirements is the responsibility of everyone involved with the creation or consumption of those processes.

Effective Quality Management also has both current and future components, that is, understanding today's customer desires and anticipating future customer desires. Because value and satisfaction may be influenced by many factors throughout a customer's overall experience with an organization, improvement efforts must do more than reduce defects, errors, and complaints or merely meet specifications. They must address the product and service characteristics that meet basic customer requirements as well as those features and characteristics that differentiate the organization from its competitors. Such differentiation may be based on innovative offerings, combinations of product and service offerings, multiple access mechanisms, user-friendly processes, rapid response, or special relationships.

In order to be a quality-conscious organization, efforts must consistently focus on

- Customer retention and loyalty
- Continually improving performance
- Market share gain and growth
- Sustaining unity among employees and suppliers
- Changing and emerging customer and market requirements
- Factors that drive customer engagement
- Changes to the marketplace

Knowledge Management

Organizations and their employees are faced with a continuous stream of information from various sources, including conversations, documents, meetings, e-mails, books, and websites. With such a surplus of content, many critical pieces of knowledge such as lessons learned, training, and best practices are often overlooked or go unused by leadership and staff within an organization. This led to the creation of a discipline known as *Knowledge Management*, or KM, in the early 1990s in order to better control and manage the vast amounts of data, information, and knowledge within an organization. Since knowledge is a fluid mix of experiences, values, contextual information, and professional insight, making these knowledge assets readily available to the right people within an organization can be a tremendous benefit. KM is the process of capturing, distributing, and effectively using knowledge to enhance organizational performance. The overarching goal of KM is to create organizational conditions that enable and promote the creation, sharing, and use of valuable knowledge so that an organization can create enhanced economic and social value for its customers and community.

KM is comprised of the following three knowledge types:

- *Explicit Knowledge:* Knowledge that has been clearly expressed and communicated as words, numbers, codes, or formulas. It is easy to communicate, store, and distribute. Sources typically include books, websites, and other visual or verbal communications.

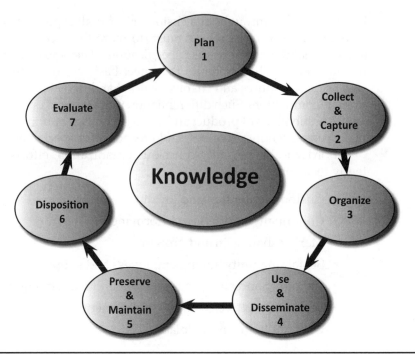

FIGURE 6-17   Knowledge Management LifeCycle

- *Tacit Knowledge:* Knowledge that is unwritten, unspoken, or hidden. This information is usually based on an employee's experiences, insights, and observations. It is communicated through discussion, association, or activity with other coworkers or leaders.
- *Embedded Knowledge:* Knowledge that is contained within organizational processes, products, culture, artifacts, or structures and is often a part of routine tasks.

Figure 6-17 depicts the seven stages of the KM Life Cycle.

NOTE   *The KM life cycle is repetitive in nature, as are most Process Improvement methods, and is an ongoing activity.*

Benefits of Knowledge Management

Knowledge Management helps companies and individuals

- Improve operational effectiveness and efficiency
- Harvest and transfer knowledge across the organization

- Foster innovation and cultural change
- Capture and transfer critical knowledge from retiring experts to new employees
- Bridge knowledge gaps that exist between formal classroom training and field experience
- Enhance knowledge about customers and the organization's environmental factors
- Improve organizational performance and profitability
- Manage organizational risk
- Use knowledge as a competitive advantage to drive innovation
- Increase retention of experienced, knowledgeable employees

Knowledge Management Tools and Practices

Several KM practices help organizations share and disseminate knowledge. These include

- *Communities of Practice:* These are groups of individuals who share a concern or a passion for a particular topic and formally gather to communicate knowledge and learn from one another. General meetings can be formal or informal and occur at a frequency agreed upon by the group. Members recognize the collective value of sharing information and have established norms of trust and cooperation.

- *Document Repositories:* These are databases of collected knowledge assets that are organized in order to facilitate searching, browsing, and retrieval of corporate knowledge artifacts. Document repositories often contain lessons-learned documents, Process Improvement artifacts, standards and best practices, project planning documents, and templates.

- *After-Action Reviews:* These are facilitated meetings that occur after significant and intensive performance or project events in which process operators or project team members collectively analyze and determine the root cause of a particular issue.

- *Retrospectives:* These are facilitated meetings that occur at the conclusion of a Process Improvement project or iteratively throughout the project. At these meetings process operators and project team members collectively analyze and reflect on what activities went well in a particular project or time period, what components could be improved, and what successes can be incorporated into future iterations or improvement initiatives.

Other Knowledge Management instruments include

- Knowledge fairs
- Knowledge repositories
- Social Wikis, blogs, or e-collaboration sites
- Expertise directories
- Intra-project learning
- Inter-project learning

Chapter Summary

Chapter 6 presented a framework which process-focused organizations could use to better manage Process Improvement efforts and Strategic Objectives. It also outlined the various methods used to manage Process Improvements projects and described the importance of incorporating agility into all facets of performance management. From this chapter, we also learned that:

- A framework describes the rationale of how an organization creates, delivers, and captures value.
- There are two types of environmental influences: External Driving Forces and Internal Driving Forces.
- A Super System Map is a pictorial view of an organization's relationship with its business environment and includes the following six principal components:
  - Organization
  - Customers
  - Suppliers
  - Competitors
  - Environmental influences
  - Shareholders/Stakeholders (if applicable)
- Strategic Planning is an organization's process for defining its strategy or direction and for making decisions on allocating its resources to pursue this strategy, including its capital and people.
- Process Management is the ensemble of activities for planning, engineering, improving, and monitoring an organization's processes.
- There are several widely used methods for managing Process Improvement projects including Lean Six Sigma and Rummler–Brache.

- Many of the principles found in the Agile Manifesto can be repurposed for Process Improvement.

- Performance Management is the ongoing practice of ascertaining how well or, how poorly, an organization and its processes are performing.

- Workforce Management describes an organization's system for developing and assessing the engagement of its workforce, with the aim of enabling and encouraging all members to contribute effectively and to the best of their ability.

- Quality is known as a measurement of excellence or as a state of being free from deficiencies or significant variations against stated or implied customer requirements.

- Knowledge Management is the process of capturing, distributing, and effectively using knowledge to enhance organizational performance.

Chapter Preview

Now that the components of the Process Improvement Framework have been discussed, the next step is to learn the various benefits of centralizing the management of Process Improvement activities into a Process Improvement Organization. Chapter 7 provides an overview of the Process Improvement Organization and its various services. Various Process Improvement controls are also discussed.

CHAPTER 7

The Process Improvement Organization (PIO)

Process Improvement Organizations (PIOs) are a core component of any company's quality system. The fundamental purpose of establishing a PIO is to increase an organization's efficiency and the quality of its products and services in order to better manage stakeholder and customer demands. As such, it is prudent for any organization to follow a disciplined approach to improvement activities in order to respond effectively and rapidly to stakeholder needs. The PIO is a service-driven organization within a corporation and is responsible for providing Process Improvement oversight and ensuring that all Process Improvement activities are accurately aligned with stakeholder needs and expectations. This chapter is a guideline and companion for Process Improvement professionals as they build out a Process Improvement function within an organization. It outlines the various roles, behaviors, services, management controls, and techniques used to deliver value to stakeholders. Furthermore, this chapter describes the structure and makeup of a Process Improvement Organization (PIO).

This chapter is organized around the following topics:

- *Charter of a PIO:* What are the primary objectives of a PIO and what is the department's mandate?

- *Process Improvement Services:* What services does the PIO provide? What are the estimated timelines associated with each service?

- *Roles and Responsibilities:* What are the fundamental roles that make up a PIO?

- *Process Improvement Governance Structure:* How are Process Improvement projects governed and what role does the PIO play in this process?

- *Department Controls:* What controls are in place to ensure a consistent delivery experience for Process Improvement stakeholders?

Charter of a Process Improvement Organization

Process Improvement Organizations (PIOs) are assigned various responsibilities associated with the centralized and coordinated management of Process Improvement efforts and projects across an enterprise. A PIO's vision is to continually improve all aspects of the enterprise's environment (process, structure, workforce, technology) in order to improve organizational performance. This vision is put into action through various Process Improvement programs along with a focus on process stewardship, activities to eliminate waste, a commitment to building customer value, and a dedication to sustaining enterprise operations. Ultimately, the PIO is responsible for optimizing processes to achieve more efficient results while ensuring alignment with an organization's strategic objectives. Using disciplined and systematic approaches such as Lean Six Sigma, Kaizen, and Rummler–Brache, the PIO strives to improve operations on behalf of customers and shareholders.

Specifically, the PIO focuses on five distinct areas (Figure 7-1):

- *Enterprise understanding:* Ensuring proper business architecture exists and all processes are documented and understood. Managing any changes to an organization's Process Ecosystem and considering the complete enterprise before any process changes are implemented.

The Process Improvement Organization focuses on five distinct areas:

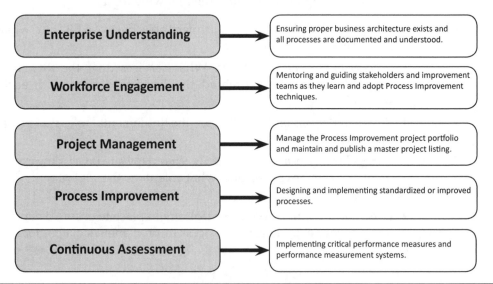

FIGURE 7-1   Process Improvement Organization Focus Areas

- *Workforce engagement:* Mentoring and guiding stakeholders and improvement teams as they learn and adopt Process Improvement techniques. Providing training to all levels of an organization so that supervisors, middle managers, and executives understand their roles in making Process Improvement a success. Educating the organization on Process Improvement activities and publicly praising accomplishments to reinforce improvement actions. Ensuring employees and process operators are considered in any process decision and proper change management activities occur.

- *Project management:* Managing the Process Improvement project portfolio and maintaining and publishing a master project listing. Ensuring appropriate methods and tools are used in Process Improvement efforts. Participating in governance activities and serving as an honest broker on all issues brought forward to the PIO by business stakeholders.

- *Process Improvement:* Designing and implementing standardized or improved processes. Reducing unprofitable or wasteful activities such as defects, overproduction, transportation, duplication, waiting, and unnecessary motion or processing. Ensuring Process Improvement activities align with strategic objectives.

- *Continuous assessment:* Implementing critical performance measures and performance measurement systems. Ensuring that proper monitoring of performance occurs and that any variation is assessed and improved if necessary. Serving as the organization's authority on Process Improvement Practices by setting the standard, providing the tools and templates, and then being a resident advocate and model for good Process Improvement practice.

Process Improvement Services

A PIO provides a number of professional services to assist organizations in achieving process excellence. These services range from directly managing Process Improvement projects, to implementing and modeling business process architectures, to simply providing Process Improvement support to business customers by way of guidance and education. These activities form the backbone of the Process Improvement group and enable organizations to improve performance in a coordinated, centralized, and disciplined manner. This section provides a comprehensive overview of the services offered by a PIO along with a description of the service, an outline of standard activities, and any key deliverables. Estimated time frames or service-level agreements for fulfilling each service are also provided.

NOTE Timelines for the various Process Improvement services vary among organizations depending on a variety of factors such as the complexity of the process in question; the nature, quantity, and magnitude of process disconnects; the state of the process (a documented or undocumented process); the ability to dedicate resources to the effort (an item which can drastically alter time lines) and whether there are any other competing or conflicting Process Improvement goals within the organization. These estimates are provided as a baseline for discussion. Ultimately, whether a Process Improvement endeavor is multiday, multiweek, or multimonth, there is always a benefit to conducting the effort. Incorporating agility into traditional Process Improvement life cycles can help improve the speed of value delivery.

Project Intake

The *Project Intake* process exists in order to ensure all Process Improvement service requests are properly assessed for prioritization, resourcing, and execution. This service includes formal triaging/investigation to properly define the problem statement of the request, formally documenting the request, and ensuring adequate preparation for governance and executive or steering committee review.

For this service, the PIO will

- Register the Process Improvement idea into a master project list or repository
- Investigate the improvement idea
- Document the problem, business case, and proposed benefits of the proposed improvement
- Ensure strategic alignment of the idea with departmental Hoshin Kanri and strategic goals
- Estimate the effort required to execute the proposed improvement
- Select an appropriate project method for delivery
- Obtain approval to proceed with the improvement request and estimate a timeline for completion

A typical baseline for completing analysis of a Process Improvement idea is as follows:

- Low-complexity request: 5 days
- Medium-complexity request: 10 days
- High-complexity request: 15 days

Process Improvement Project

A *Process Improvement Project* is a large-scale initiative that involves complex processes that include multiple stakeholders across the enterprise, multiple inputs and outputs, and solutions that often require significant time and investment to implement. This service offering includes structured sets of activities designed to improve the performance of business processes in order to fulfill organizational goals. Throughout the project life cycle, participants verify the current state model of a process (As-Is) and propose designs for the future state model (To-Be). It provides diagnostic information on issues to be improved and opportunities for beneficial change. Moreover, Process Improvement resources provide guidance and delivery execution throughout the various improvement phases.

For this service, the PIO will

- Produce the Project Charter, Project Plan, and As-Is Process map
- Define and collect supporting data and measurements
- Perform Root Cause Analysis (RCA)
- Identify potential solution(s) and create a To-Be Process Map
- Confirm the proposed solution fits process architecture
- Execute identified quick wins
- Engage technology teams to assess any proposed process automation solutions
- Obtain approvals from management for solution selection and prioritization
- Create a Change Management plan where applicable for Process-related changes
- Complete and execute a training plan for process change adoption
- Update the Process Ecosystem and associated process components such as procedures
- Identify measures and create a process control dashboard
- Prepare for and present to stakeholders and management at Tollgate Reviews (these occur at various stages of a Project)

Typical deliverables for a Process Improvement Project include

- Measured improvements to business processes that meet customer requirements and organizational objectives
- Documented Voice-of-Customer, Critical-to-Quality, and success measures
- Reports and/or dashboards that measure business performance
- Trained workforce on new processes and procedures
- Implemented, validated, approved, and documented solutions including process maps and procedures
- Response plan for resolution of potential process issues

Common timelines for executing Process Improvement projects are as follows:

- Low-complexity project: 40 days
- Medium-complexity project: 80 days
- High-complexity project: 120 days

Figure 7-2 outlines the activities typically contained in a Process Improvement project using the Lean Six Sigma DMAIC (Define-Measure-Analyze-Improve-Control) method.

Task #	Task Name
	Define Phase
1	Identify Stakeholders
2	Write Project Charter
3	Obtain approval through intake and governance to proceed with the improvement initiative
4	Establish a Project Plan
5	Determine and map as-is processes
6	Identify quick wins
7	Understand the decision making landscape that impacts the project (e.g., Create a stakeholder management plan)
8	Create a plan to communicate project progress
9	Identify Voice of the Customer (VOC), translate into Critical to Quality (CTQ) metrics and define success measures
10	*Deliverable: Documented VOC, CTQ and success measures*
11	**Milestone: Tollgate review with management**
12	Modify Define Phase deliverables based on tollgate review, if necessary
13	**Milestone: Define Phase Complete**
	Measure Phase
14	Identify measures
15	Create a data collection plan
16	Collect, interpret, describe and display data
17	*Deliverable: Report(s) and/or dashboard(s) that measures business performance*
18	Identify baseline performance from which improvements will be compared
19	**Milestone: Tollgate review with management**
20	Modify Measure Phase deliverables based on tollgate review, if necessary
21	**Milestone: Measure Phase Complete**
	Analyze Phase
22	Perform root cause analysis
23	Conduct process analysis to determine non-value added steps
24	Perform data and statistical analysis
25	*Deliverable: Root Cause Analysis*
26	**Milestone: Tollgate review with management**
27	Modify Analyze Phase deliverables based on tollgate review, if necessary
28	**Milestone: Analyze Phase Completed**

Figure 7-2 Process Improvement Project Activities

Task #	Task Name
	Improve Phase
29	Identify and define potential solution(s)
30	Determine and map potential to-be processes
31	Determine the impact of the solution(s) on Process Ecosystem and Architecture
32	Conduct technology, risk, compliance and dependency assessment on proposed solution(s)
33	Determine cost/benefit analysis of potential solution(s)
34	Select the solution(s) to proceed with
35	Prioritize solution(s) for iterative development and deployment
36	Obtain management approval for solution selection and prioritization
37	Create and execute process Change Management Plan
38	Complete an implementation and training plan
39	Update process library and job procedures
40	Train operators and management
41	*Deliverable: Trained operators and management about the new procedure(s)*
42	Test process and procedures for the solution
43	Obtain change management approval for process changes
44	Implement process changes
45	*Deliverable: Implemented, validated, approved and documented solution(s) including process maps and procedures*
46	**Milestone: Tollgate review with management**
47	Modify Improve Phase deliverables based on tollgate review, if necessary
48	**Milestone: Improve Phase Completed**
	Control Phase
49	Identify measures for process control
50	Design process control dashboard
51	*Deliverable: Report(s) and/or dashboard(s) to measure ongoing business performance*
52	Create a response plan for process issue resolution
53	*Deliverable: Response plan for resolution of potential process issues*
54	Complete a process transfer plan for operators and management
55	Communicate results to stakeholders and management
56	Evaluate outstanding or new Process Improvement ideas discovered during the project
57	Submit new continuous improvement ideas through Continuous Improvement Intake
58	Conduct Project Retrospective
59	**Milestone: Tollgate review with management**
60	Modify Control Phase deliverables based on tollgate review, if necessary
61	Close the project
62	**Milestone: Control Phase Completed**

FIGURE 7-2   (Continued)

Kaizen Event

A *Kaizen Event* is a focused effort to make rapid improvements over a brief period of time. Cross-functional teams are assembled for three- to five-day periods in order to focus on improving a particular process. Kaizen Events are

- A short burst of intense activity and effort (three to five days)
- Biased toward action over analysis to achieve improvements in a short time
- Focused on defining activities, improving supplier/customer connections, and achieving flow
- Driven to resolve a specific problem or achieve a specific goal
- Committed to a specific area or process
- Guided with daily reviews of progress

With this service offering, the PIO provides guidance and assistance throughout the planning and facilitation of Kaizen events as well as any solution deployment activities post Kaizen.

For this service, the PIO will

- Identify the needed participants
- Gather applicable/available performance data
- Schedule the event and prepare the agenda
- Map the As-Is process
- Identify process disconnects and determine root causes
- Design and propose solutions
- Confirm that the proposed solutions fit the process architecture
- Obtain change management approvals
- Implement the process changes
- Update process map(s), procedures, and documentation
- Analyze the results
- Summarize and present what was learned to management
- Develop a Process Monitoring plan

Typical deliverables for a Process Improvement Project include

- Implemented, validated, approved, and documented solution(s) including process maps and procedures
- Report(s) and/or dashboard(s) to measure ongoing business performance

Kaizen Planning, Facilitation, and Implementation are usually completed within 15 days as follows:

- 5 days for Kaizen planning
- 5 days for Kaizen event an implementation
- 5 days for post implementation activities

Figure 7-3 outlines the activities typically contained in the Kaizen event service.

Process Change Request: Just Do It

This service offering provides assistance with the implementation of a Just-Do-It process change. Just-Do-It changes require little effort to implement, and root causes and appropriate solutions are well known. Examples of

Task #	Task Name
	Plan Kaizen
1	Determine Kaizen objectives and complete Initiative Proposal Document
2	Obtain management approval through intake and governance
3	Identify Kaizen participants
4	Gather available performance data
5	Schedule Kaizen event
6	**Milestone: Plan Phase Completed**
	Kaizen Event
7	Map as-is process
8	Determine the problem and root cause
9	Identify process disconnects and propose solutions
10	Solicit buy-in from all parties related to the process
	Post Kaizen
11	Obtain change management approval
12	Implement process changes
13	Update process maps, procedures and document solution(s)
14	*Deliverable: Implemented, validated, approved and documented solution(s) including process maps and procedures*
15	**Milestone: Do Phase Complete**
16	Analyze results
17	Summarize what was learned
18	Present results to management and the organization
19	**Milestone: Check Phase Complete**
20	Develop a monitoring plan
21	Transition process change to operational team
22	*Deliverable: Report(s) and/or dashboard(s) to measure ongoing business performance*
23	Prepare a plan for another Kaizen event, if necessary
24	Implement improvements on a wider scale, if appropriate
25	**Milestone: Act Phase Complete**

FIGURE 7-3 Kaizen Event Activities

Task #	Task Name
	Just Do It
1	Determine Process Change objectives
2	Validate the problem still exists
3	Confirm the solution fits
4	Prepare Change Management Plan
5	Obtain change management and architecture approval
6	Implement the solution
7	*Deliverable: Implemented, validated, approved and documented solution(s) including process maps and procedures*
8	Update any related process artifacts
9	Communicate results to stakeholders and management
10	**Milestone: Just Do It Completed**

FIGURE 7-4 Just-Do-It Activities

Just-Do-It scenarios include simply transferring a series of activities to a new department, retiring a particular activity or series of activities, and renaming or rebranding various process attributes.

For this service, the PIO will

- Validate the problem or concern
- Confirm that the proposed solution fits process architecture
- Prepare a change management plan
- Update process map and procedures
- Obtain approvals for the plan
- Implement the solution
- Update any documentation, reports, or other process artifacts
- Communicate the results to stakeholders and management

Just-Do-It Scenarios are typically completed in less than five days.

Figure 7-4 outlines the activities typically contained in a Just-Do-It or Process Change Request scenario.

Process Mapping (As-Is or To-Be)

Process Mapping is essential for understanding all aspects of an organization's workflows and processes, including departments, activities, owners, and hand-off points. This documentation assists with standardization activities and helps ensure tasks are executed in a consistent manner among process operators. In addition, it is a helpful tool when disseminating information and training employees; and in many cases, it may be required by regulatory entities. With this service offering, the PIO assists stakeholders with mapping either the As-Is or To-Be state of a named process; documenting any related process components such as associated business rules, measures, dashboards, procedures, or systems; and properly deploying the process into the organization's Process Ecosystem.

For this service, the PIO will

- Identify the start and end points of the process in question
- Identify the operators and Subject Matter Experts (SMEs)
- Schedule and facilitate workshops/interviews
- Document the process map
- Document the Process Profile
- Document any Process Components
- Update the Process Ecosystem and Process Repository
- Validate the process map and profile with stakeholders
- Conduct training with stakeholders and SMEs, as needed
- Communicate the results to stakeholders and management
- Conduct retrospective

Typical deliverables for a Process Mapping engagement include

- Documented As-Is and/or To-Be Process
- Process Profile
- Implemented, validated, approved, and documented process components
- Updated view in the centralized process repository

Common service-level agreements for a Process Mapping engagement are as follows:

- Low-complexity process: 5 days
- Medium-complexity process: 10 days
- High-complexity process: 15 days

Figure 7-5 outlines the activities typically contained in a Process Mapping Exercise.

Root Cause Analysis

There are times when there are symptoms or indicators of a process problem; however, the validity or cause of the performance fluctuation is unknown. *Root Cause Analysis* (RCA) is a problem-solving method that is used to identify the root causes of process fluctuations or problems that cause operating events that affect performance. With this service offering, the PIO assists process stakeholders with completing formal

Task #	Task Name
	Process Mapping
1	Determine Process Mapping objectives
2	Obtain management approval through intake and governance
3	Identify operators and Subject Matter Experts (SME)
4	Identify the process, start point(s) and end point(s) to be mapped
5	Schedule workshops/interviews
6	Facilitate workshops/interviews with operators and SME's
7	Document Process Map
8	Document Process Profile
9	Validate Process Map and Process Profile with stakeholders and management
10	Update Process Ecosystem as needed
11	*Deliverable: Documented As-Is and/or To-Be Process including Process Profile, if necessary*
12	Communicate results to stakeholders and management
13	**Milestone: Process Mapping Complete**

Figure 7-5   Process Mapping Activities

RCA of process issues or performance disturbances and facilitating the design of any needed improvements.

For this service, the PIO will

- Identify appropriate analysis methods to use and gather data
- Schedule workshops and interviews to determine the root cause of an issue
- Facilitate problem identification and conduct RCA
- Prepare and document the findings
- Recommend potential solutions and methods for deployment
- Communicate results to stakeholders and management

In many cases, process breakdowns require immediate attention and root cause efforts should be conducted within hours of any incident. However, for noncritical events, common service-level agreements for Root Cause Engagements are as follows:

- Low complexity: 1 day
- High complexity: 3 days

Figure 7-6 outlines the activities typically contained in RCA efforts.

Consulting

Continuous Improvement can provide various types of consultation services for business customers within an organization. Services include providing advice regarding quality control, process mapping standards, improvement methodologies, industry best practices, strategy alignment,

Task #	Task Name
	Root Cause Analysis
1	Determine Root Cause Analysis objectives
2	Identify analysis methods to use (Affinity, Fishbone, Lean Process, Data, etc.)
3	Gather data
4	Identify operators and SME's
5	Schedule workshop(s)/interview(s) for determination of root cause
6	Conduct root cause analysis
7	Prepare and document findings
8	Recommend solution(s) to resolve root causes
9	*Deliverable: Root Cause Analysis Document including analysis artifacts and recommendations for improvement*
10	Communicate results to stakeholders and management
11	**Milestone: Root Cause Analysis Complete**

FIGURE 7-6   Root Cause Analysis Activities

as well as providing support for System Development activities in order to prevent needless development and automation.

For this service, the PIO will

- Identify artifacts to be delivered based on consultation needs
- Identify any needed stakeholders, operators, and SMEs
- Schedule and facilitate workshops/interviews
- Prepare and validate deliverables with stakeholders, operators, and SMEs
- Communicate results to stakeholders and management

Most consultations can be completed within a few hours by scheduling or attending various project or department meetings. Figure 7-7 outlines the activities typically contained in a Process Improvement consulting engagement.

Task #	Task Name
	Process Consulting
1	Determine Consulting objectives
2	Identify artifacts to be delivered
3	Identify stakeholders, operators and SME's
4	Schedule workshop(s)/interview(s)
5	Prepare deliverables
6	Validate deliverables with stakeholders, operators and SME's
7	*Deliverable: Completed artifacts as determined by the request, if necessary*
8	Communicate results to stakeholders and management
9	**Milestone: Consulting Services Completed**

FIGURE 7-7   Process Improvement Consulting Activities

Process Training

The PIO is considered a transformation group and as such is often called upon to collaborate with business SMEs to conduct training sessions with process operators prior to implementing process changes. The PIO is also responsible for indoctrinating process culture and Process Improvement practices into the organization it supports. This service can include coaching, mentoring, or, in many cases, training various departments on business process changes, methodologies, and best practices.

For this service, the PIO will

- Create training curriculum and schedules and prepare necessary collateral material
- Schedule and coordinate training sessions
- Train stakeholders, operators, and/or SMEs
- Determine metrics to measure the net benefit of training activities

Common service-level agreements for Process Training engagements are as follows:

- Low-complexity training session: 5 days
- High-complexity training sessions: 10 days

NOTE *If existing training assets can be reused, training can often be executed in less than 10 days even for high-complexity training arrangements.*

Figure 7-8 outlines the activities typically contained in a Process Improvement training engagement.

Task #	Task Name
	Process Training
1	Identify processes to be trained
2	Create training curriculum, schedule and prepare necessary collateral
3	*Deliverable: Schedule outlining all processes to be trained and related activities*
4	Validate schedule with stakeholders, SMEs
5	Schedule training session(s) with stakeholders and SMEs
6	Conduct training session(s)
7	Conduct initial training review with stakeholders, SMEs to discuss results
8	Complete quality assessment of any transactions processed by stakeholder/SMEs post training to ensure appropriate quality levels
9	*Deliverable: Trained Process Operators and Organizational Staff*
10	**Milestone: Representative released into full production**

FIGURE 7-8   Process Training Activities

Quality and Performance Assessments

Quality Assurance is a service provided by the PIO in order to evaluate operational performance, including the delivery of services, quality of products provided to consumers or customers, and performance of any human resources involved in process execution. The PIO conducts assessments of key business processes in order to ensure transactions are meeting quality benchmarks, thus ensuring that customers receive precisely what is requested. This service helps monitor results in specific process areas and assists with identifying potential areas of concentration for Process Improvement initiatives. Quality Assessments can also assist with employee performance management.

For this service, the PIO will

- Determine critical processes for assessment
- Determine appropriate sample sizes and assess transactions
- Create audit notes to be used to coach process operators
- Perform trend analysis to identify areas of focus for Process Improvement
- Provide feedback for process operators to reduce defect rates
- Ensure performance meets regulatory compliance requirements

Conducting quality assessments on organizational processes should be a continual effort in order to ensure consistent and high-quality outcomes. Figure 7-9 outlines the activities typically contained in Performance Assessment and Evaluation efforts.

Other Process Improvement Services

Other improvement services may include performing various activities related to process, governance, or project management as well as:

Task #	Task Name
1	Determine critical process outputs
2	Assess transactions according to appropriate sample size
3	Create case notes to be used to coach process operators
4	Prepare deliverables
5	*Deliverable: Quality assessment feedback to the process operators*
6	Identify metrics to monitor trends in quality of process inputs and outputs
7	*Deliverable: Trend analysis to identify areas of focus for Continuous Improvement*
8	**Milestone: Daily quality averages and representative feedback delivered**

FIGURE 7-9 Process Performance Assessment Activities

Organizational Restructuring

Process Improvement organizations often assist with and are an integral part of organizational restructuring activities. *Restructuring* is a term used to describe the act of reorganizing the structure of a company (Process, People, and Technology) for the purpose of making it more profitable or more efficient. Other reasons for restructuring include organizational ownership changes, responses to process shifts or fluctuations in performance, or major organizational events such as buyouts or mergers. When major organizational events occur, processes should be realigned to meet the organization's new mandate. Likewise, when organizational processes change, structures, roles, and functions should be realigned with the new process objectives. In many organizations, a common misstep when restructuring is to overlook the change management activities required to ensure employees understand how to operate in the new environment, ultimately undermining effectiveness and previously stated performance targets. The best way to align or realign an organizational structure is to have full visibility and understanding of its processes in order to suggest the most optimal realignment of functions to support the proposed changes. The PIO can assist with designing changes to processes and related attributes as well as facilitate any change management activities to ensure proper alignment at all levels within the organization. The primary assets used in this service are the Process-Oriented Architecture (POA) framework and the Process Ecosystem as they provide both the guidance and visibility needed to properly analyze, design, and manage restructuring changes.

Benchmarking

Benchmarking is the process of comparing business processes and performance metrics with those that demonstrate superior performance in the industry or from other industries. Dimensions typically measured are quality, time, and cost. When benchmarking organizations identify the best firms in their industry or in another industry where similar processes exist, they compare the results and processes of those studied to their own results and processes. In this way, they learn how well the targets perform and, more importantly, the business processes that explain why these firms are successful. The PIO can use the Process Ecosystem to identify key processes and measures and facilitate the comparative analysis to assist business partners with establishing improvement targets.

Roles and Responsibilities

For a PIO to work at optimum efficiency, each member of the group must understand their roles and responsibilities. Likewise, the PIO's business customers must also understand the differences and accountabilities of the individuals being assigned to support their Process Improvement efforts. Well-defined roles and responsibilities help build consistency in

Process Improvement delivery and result in an economy of repetition that brings a sense of predictability and assurance to business stakeholders. It is important to recognize that there are three common roles that most Process Improvement organizations contain in some form. In its simplest form, a PIO contains the following roles:

- Process Improvement Coordinator
- Process Improvement Manager
- Process Improvement Architect

Process Improvement Coordinator

The Process Improvement Coordinator is responsible for assisting the PIO with all aspects of support including scheduling, facilitating/coordinating meetings, completing process documentation, and administering any Process Improvement Management systems that may be in place. The Coordinator is generally accountable for maintaining any procedural aspects of the Process Improvement group and serves as the designated point of contact for all Process Improvement issues and questions.

The overall aim of the Process Improvement Coordinator's role is to support Process Improvement efforts; however, the Coordinator may be required to facilitate small Process Improvement initiatives and change requests. The coordinator helps build and maintain the organization's process library and documentation repository to ensure that tools and templates are standardized and utilized. The Coordinator also ensures that repository permissions are in place, changes to processes are communicated, and any standards that are required for use are documented and distributed. An individual who holds this position is confident, positive, dynamic, an excellent communicator, a self-starter who focuses on quality, and has attention to detail yet still sees the big picture.

Other duties include

- Tracking service requests and actions and capturing progress
- Generating status and summary reports
- Recording as-is and should-be process maps and associated collateral
- Generating process documentation and templates
- Assisting with maintenance and updating of process artifacts
- Working closely with cross-functional departments to ensure requirements are not missed
- Assisting with promoting and maintaining process standards
- Assisting with the planning and organization of multiple process engagements

- Managing short-duration process projects/complex tasks/RCA efforts
- Assisting other members of the department as required
- Ensuring that all Process Improvement initiatives follow department standards and controls and any artifacts are audit ready
- Executing jobs to professional standards using industry methods, tools, templates, and facilitation techniques
- Supporting the organization in all aspects related to process awareness
- Providing recommendations for continuous improvement

Other titles for this role include Process Coordinator, Process Analysis Coordinator, and Process Engineering Coordinator.

Process Improvement Manager

The Process Improvement Manager is primarily responsible for developing an organization's capability by teaching Process Improvement skills and managing Process Improvement projects. The Manager is involved with various organization-wide cross-functional initiatives to enhance performance and effectiveness and is ultimately accountable for managing all phases of improvement efforts including process discovery, definition, design, simulation, implementation, and monitoring. The Process Improvement Manager ensures that any improvement effort is managed in a disciplined and structured fashion and that all efforts align with the organization's strategic objectives.

In particular, the Process Improvement Manager assists organizations with building process awareness, establishing a culture of continuous improvement, and developing process solutions that are more robust, customer centric, and lean in design. The Manager is considered an expert in Process Improvement methods and is accountable for the delivery of all process components including process maps, Key Process Indicators (KPIs), Service Level Agreements (SLAs)/Organization Level Agreements (OLAs), procedures, and dashboards. The Manager uses process management systems to create process models that can be simulated, analyzed, and executed by business stakeholders and he or she also facilitates the various workshops required to define, measure, analyze, and improve processes.

In this position, the Process Improvement Manager works with Process Improvement Architects as well as various business units and stakeholders within an organization to define the enterprise process backbone (the critical core processes), their relationships with all external entities, other enterprise value streams, and the events that trigger instantiation. This is done while executing any improvements against this structure using effective project and program management leadership.

Ultimately, the Process Improvement Manager's overall role is to facilitate and manage all Process Improvement initiatives and ensure all proposed changes to an organization's processes align with both business strategy and business architecture requirements. An individual in the Process Improvement Manager's role is confident, positive, dynamic, an excellent communicator, a self-starter who focuses on quality, and has attention to detail but still sees the big picture.

Other duties include

- Providing leadership and expertise throughout the design and improvement of processes
- Generating status and summary reports
- Modeling and mapping as-is and should-be processes and associated collateral
- Creating process documentation and templates
- Performing RCA and providing recommendations for improvement
- Ensuring that all Process Improvement initiatives follow department standards and controls and that any artifacts are audit ready
- Executing jobs to professional standards using industry methods, tools, templates, and facilitation techniques
- Mapping and Modeling business processes
- Leading large-scale transformations of complex business processes
- Designing and organizing workshops, facilitating group sessions, and working with teams to gain crucial input from key contributors
- Leading teams of SMEs through all phases of process redesign
- Communicating and executing change management activities at all levels of the organization
- Conceptualizing solutions that address customer opportunities and issues
- Understanding the latest process architecture, technology, solutions, and industry trends
- Meeting with business and technical leaders to identify and scope opportunities, outline potential value and return on investment (ROI), and identify risks and constraints
- Acting as a mentor or coach to business stakeholders by providing training and guidance
- Analyzing data for trends and recommending possible improvements and solutions
- Project Managing process improvement projects and ensuring goals are achieved on time, on budget, and within scope

Other titles for this role include Process Engineer, Workflow Engineer, Business Process Modeler, and Process Analyst.

Process Improvement Architect

A Process Improvement Architect typically performs the task of architecting and designing an organization's processes and related components. He or she is responsible for executing and maintaining POA principles and building the organizational ecosystem model needed to support the organization's operations. A Process Architect works as a change agent with business stakeholders and plays a key role in shaping and fostering continuous improvement and business transformation initiatives. The Architect has deep business knowledge as well as broad technology knowledge, combined with large-scale, cross-functional process expertise and skills in industry methodologies and standards such as Capability Maturity Model Integration (CMMI), International Organization for Standardization (ISO) 9001, and Lean and Six Sigma. The Process Architect has a rare combination of business domain knowledge, process experience, transformation talents, methodology skills, and a winning personality that helps with communication and business change management. A Business Process Architect must not only see the big picture when looking across multiple process improvement initiatives, he or she must also have business strategy talents, wide-ranging process discipline skills, architecture knowledge, and technology know-how.

The Process Architect also analyzes the organization's activities and makes recommendations pertaining to the business processes in each area, in addition to relevant and timely corrections to process components and the structuring of business information. This person illustrates the alignment, or lack thereof, between strategic goals and key business decisions regarding products and services. This individual works to develop an integrated view of the organization, using a repeatable approach, cohesive framework, and available industry standard techniques. The Process Architect plays a role in driving program teams to achieve consensus about the concept and design of a solution and he or she works with leadership on the right approach. The Architect is also responsible for ensuring traceability of features back to requirements and scenarios, so that all features can be seen to provide business value and the impact of any feature changes can be assessed.

In addition, the Process Architect works with individuals at every level of the organization to solicit strategic imperatives from senior leaders and executives as well as supporting business unit managers as they leverage business architecture artifacts to execute their business processes. The Process Architect may provide direct input into the governance cycle for the funding, oversight, and realization of business projects. In that governance role, the Process Architect helps to ensure that business and information technology projects are aligned to support the achievement of key goals, that specific business scenarios are considered, and that the overall business architecture is considered. Ultimately, a Process Architect leads efforts aimed at building an effective architecture for the Process Improvement projects that make up the business change program.

Other duties include

- Developing a process architecture strategy for business units and organizations based on a situational awareness of various business scenarios and motivations
- Applying a structured process architecture approach and methodology such as the POA for capturing the key views of the organization
- Capturing the tactical and strategic business goals that provide traceability through the organization and mapping them to metrics that provide ongoing governance capability
- Enumerating, analyzing, cataloging, and suggesting improvements to the strategic, core, and support processes of the business unit
- Defining the data elements shared between the various business units in the enterprise and defining the relationships between those data elements and processes, people, systems, and other process elements
- Enumerating, analyzing, and suggesting improvements to the structural relationships of the business
- Recognizing structural issues within the organization as well as functional interdependencies and cross-silo redundancies such as role alignment, process gaps and overlaps, and business capability maturity gaps
- Applying architectural principles, methods, and tools to business challenges
- Modeling business processes using a variety of tools and techniques
- Using these models to collect, aggregate, or disaggregate complex and conflicting information about the business
- Conceptualizing solutions that address customer opportunities and issues
- Understanding the latest architecture, technology, solutions, and industry trends
- Meeting with customers, business stakeholders, and technical leaders to identify and scope the business opportunities, outline potential value and ROI, and identify risks and constraints

- Providing architectural advice, planning guidance, quality assurance, and guidance to the project team on project approach and solution

- Managing key strategic internal client relationships with project stakeholders and senior management and interacting at a high level with decision-makers both internally and externally

- Overseeing development and maintenance of critical processes, maps, and related artifacts (Process Ecosystem) using the POA approach

- Infusing expertise in process design, efficiency, and process monitoring across the organization

Other titles for this role include Process Architect, Business Architect, Enterprise Architect, and Business Process Modeler.

Process Improvement Governance

Organizations require a purposeful set of mechanisms for managing goals and ensuring processes and Process Improvement projects align with strategic objectives that meet and exceed customer expectations. Furthermore, organizations require a method to ensure that resources and capital investments are directed toward the most critical initiatives and performance-enhancing activities. In order to do this successfully, organizations must develop a governance structure to guide their operations. Creating a Governance model fosters proper decision-making and accountability and ensures that all Process Improvement efforts align with strategic goals.

The term *Governance* refers to the set of policies, roles, responsibilities, and processes that an organization establishes in order to direct, guide, and control how the organization uses its workforce, processes, and technologies to accomplish business goals. Effective governance anticipates the needs and goals of an organization's teams and business divisions and properly analyzes any changes to its people, processes, and technology. Governance also provides policies and guidelines that make the deployment of processes and technologies both manageable for process and technology teams and effective for business and operational departments. It can help protect an enterprise from security threats or noncompliance liability and can help ensure the best ROI by enforcing best practices in change management and Process Improvement efforts. Ultimately, Governance encompasses the activity or process of governing; the individuals charged with governing; the decisions made during the governing process; and the manner, method, and system by which a particular organization governs itself.

NOTE *Every organization has unique needs and goals that affect its approach to governance. No single approach fits the cultures or requirements of all organizations. For example, larger organizations require more governance than smaller ones. However, all organizations should institute some form of governance to encourage desirable behavior within an organization and outline the set of rules and checkpoints needed to effectively manage Process Improvement activities.*

Purpose of Governance

The purpose of Governance is to

- Provide a decision-making framework that is logical, robust, and repeatable to govern an organization's projects, processes, resources, and investments
- Outline the relationships between all internal and external groups involved in all initiatives
- Describe the proper flow of information to all stakeholders regarding project and process performance
- Ensure that reviews of performance occur on a regular basis
- Ensure the appropriate review of issues encountered within improvement initiatives
- Ensure that required approvals and direction for initiatives are obtained at each appropriate stage of a Process Improvement project
- Assess compliance of the completed improvements with their strategic objectives
- Ensure that investments generate real value for the organization
- Oversee changes to organizational assets such as
  - Human assets (people, skills, career paths, training, mentoring, aptitudes)
  - Financial assets (cash, investments, liabilities)
  - Physical assets (facilities, equipment, maintenance, security)
  - Intellectual property assets (copyrights, trademarks, patents)
  - Information and technology assets (processes, data, information, knowledge about customers, applications, systems)
  - Relationship assets (internal relationships, external relationships, brand, reputation, relationships with competitors and regulators)

Governance Structure

An effective Governance structure is a key component of any process-focused organization and enables enterprise-wide process improvement and transformation. Figure 7-10 outlines a baseline governance model for Process-oriented organizations to structure their Governance activities around.

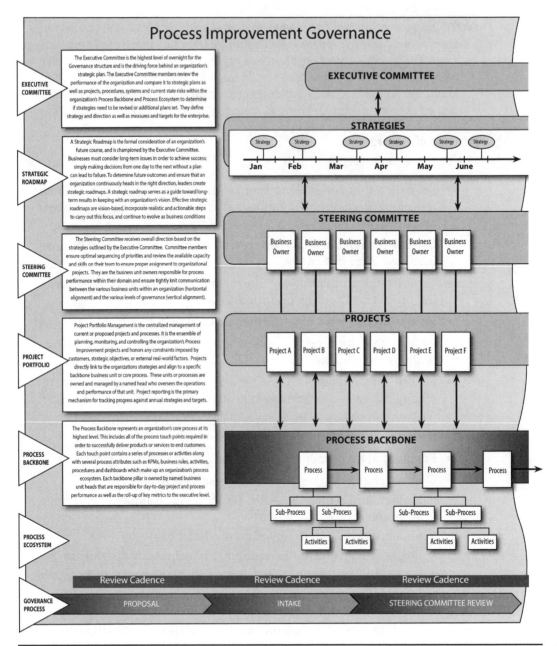

FIGURE 7-10 Process Improvement Governance Structure

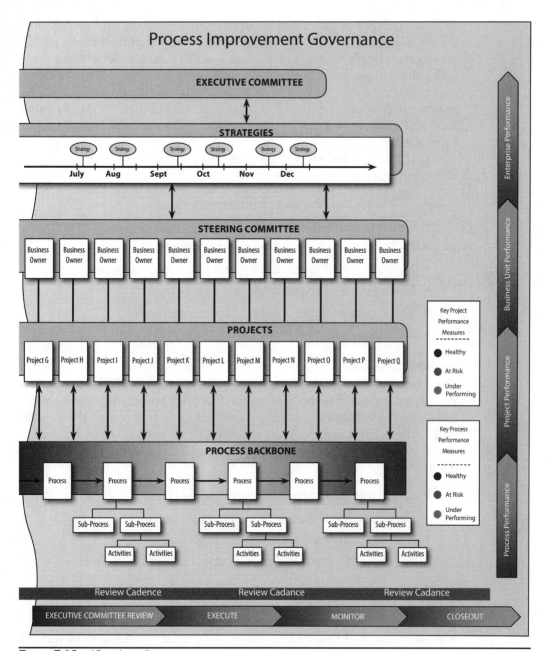

FIGURE 7-10 (Continued)

In it, the linkage of processes to departments, projects, and Key Performance Measures (KPMs) drives understanding of performance within an organization's entire ecosystem, enabling better decision-making and enhancing overall performance. The Process Improvement Governance structure ultimately assists organizations as they manage the vast number of processes and projects that exist within their enterprise portfolio and ensures strategic alignment and optimal use of resources and investments.

The Process Improvement Governance construct is comprised of the following seven parts:

- Forming an executive committee
- Creating a strategic road map
- Building management steering committees
- Managing the Process Improvement Project Portfolio
- Managing the Process Backbone and Ecosystem
- Executing the organizational governance process
- Monitoring process and organizational performance

Executive Committee

The Executive Committee is the highest level of oversight for the Governance structure and is the driving force behind an organization's strategic plan. The committee, which is comprised of senior executives from within the organization who are usually C-level executives or vice presidents, intends to achieve alignment between the following organizational levels:

- Organization and executive committee
- Executive committee and business functions
- Business functions and business processes
- Business processes and process operators and performers

The Executive Committee members review the organization's performance and compare it to strategic plans as well as projects, procedures, systems, and current state risks within the organization's Process Backbone and Process Ecosystem to determine if strategies need to be revised or additional plans set. They define strategy and direction as well as measures and targets for the enterprise.

Strategic Road Map

Strategic Planning is the process of defining strategy or direction and measures and targets and then making decisions on allocating resources to pursue this strategy, including capital and people through projects. A Strategic road map is the formal consideration of an organization's future course and is championed by the Executive Committee. Businesses must consider long-term issues in order to achieve success; simply making decisions from one day to the next without a plan can lead to failure. To determine future outcomes and ensure that an organization continuously heads in the right direction, leaders create strategic road maps. This road map serves as a guide toward long-term results in keeping with an organization's vision. Effective strategic road maps are vision based, incorporate realistic and actionable steps to carry out this focus, and continue to evolve as business conditions change. Following are the five basic steps organizations can take to craft their strategic road maps:

1. *Establish a Vision:* Describe what the results of achieving the organizational plan will look like.

2. *Define Goals:* Outline any strategic projects that must be accomplished to achieve the vision.

3. *Analyze Gaps:* Describe the current state of operations and the desired state of performance. Describe the business, technology, financial, and other factors that are required to bridge the gap.

4. *Develop a Plan:* Outline the workforce, technology, financial, and other resources, with an accompanying timeline, that are required to bridge the gap.

5. *Finalize the Road Map:* Circulate the strategic road map to relevant parties and modify it as required.

Steering Committee

The Steering Committee receives overall direction based on the strategies outlined by the Executive Committee. Steering Committee members, typically department managers and process owners, evaluate process performance within their area of responsibility and track progress of projects set forth to improve performance across the organization.

Committee members ensure optimal sequencing of priorities and review their team members' available capacity and skills to ensure proper assignment to organizational projects. They are the business unit owners responsible for process performance within their domain and ensure close communication among the organization's business units (horizontal alignment) and the various levels of governance (vertical alignment).

During scheduled Steering Committee reviews, the committee evaluates new requests along with other work that is assigned or in the queue and sequences these priorities optimally.

Project Portfolio Management

Project Portfolio Management is the centralized management of current or proposed projects and processes. It is the combination of planning, monitoring, and controlling the organization's Process Improvement projects and honors any constraints imposed by customers, strategic objectives, or external real-world factors. Projects directly link to the organization's strategies and align with a specific backbone business unit or core process. These units or processes are owned and managed by a named head who oversees that unit's operations and performance. Project reporting is the primary mechanism for tracking progress against annual strategies and targets.

Process Backbone and Ecosystem

The Process Backbone represents an organization's core process at its highest level. This includes all process touch points required to successfully deliver products or services to end customers. Each touch point contains a series of processes or activities along with several process attributes such as KPMs, business rules, activities, procedures, and dashboards that make up an organization's process ecosystem. Each backbone pillar is owned by named business unit heads who are responsible for day-to-day project and process performance as well as the roll-up of key metrics to the executive level.

Governance Process

The Governance Model includes a Governance Process to evaluate process changes, propose new project ideas, and suggest process improvements. Given that resources are finite, only the most beneficial initiatives in terms of business value should be selected for implementation. The Governance Process is divided into the following seven stages: Proposal, Intake, Steering Committee Review, Executive Committee Review, Execute, Monitor, and Closeout. Figure 7-11 outlines a basic governance process. The purpose of the governance Process is to provide a consistent and well-defined method for implementing change to an organization's processes and technology. The process focuses on delivering business value to stakeholders from the moment a request is submitted to the implementation and completion of the process or technology change.

Proposal   The driving force behind the proposal of a new project or process is the strategic plan. To foster a culture of continuous improvement, any employee may submit a project or Process Improvement proposal that will assist the organization in attaining strategic objectives. Project or Process Improvement proposals may also be derived from monitoring organizational performance at the Business Unit, project, or process level. The Executive Committee, Steering Committee, Business Leaders, Process Improvement team, or individual worker may initiate a proposal. These projects or ideas may be related to the creation or revision of processes, procedures, training, quality, compliance, reporting, or services typically offered by a PIO. Proposals are fielded by the Process Improvement team and channeled through the Intake process.

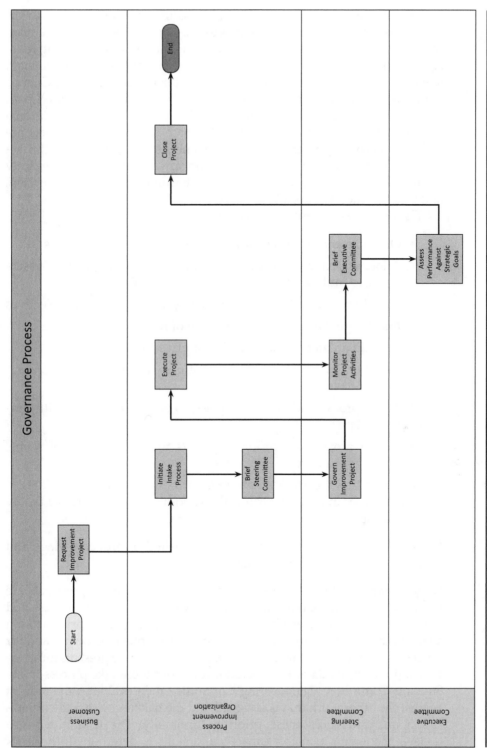

Figure 7-11 Process Improvement Governance Process

191

Intake PIOs should have a defined process to evaluate service requests for documenting and improving processes. This process allows the team to centralize and track all requests, including requests to document, redesign, or create new processes. Intake processes usually consist of two main streams: one for Change Requests and one for Project Requests. Change Requests are more frequent, simpler, and follow a more expedited path for completion and execution, while a Project Request requires additional rigor and follows a more defined path. Both paths ensure projects are delivered according to the organization's needs and priorities. Improvement opportunities sent to the Process Improvement team are regularly reviewed by a Steering Committee to determine feasibility, resourcing, and prioritization against other efforts.

The purpose of an intake process is to

- Organize and prioritize new Process Improvement requests
- Allocate resources to critical process efforts and reduce excessive resource multitasking
- Set achievable and visible time frames to meet customer objectives
- Encourage the setting and attaining of mutual goals
- Provide predictable and consistent project estimating
- Register all intake items to ensure timely investigation, prioritization, and completion
- Triage project and process requests to ensure alignment with strategic plans
- Define project and process requests to ensure clarity of scope, timelines, resources, and budget
- Centralize and track all requests for new projects and updates to processes in order to ensure all proposals are managed in a timely manner

Figure 7-12 outlines the typical activities needed to properly triage and evaluate Process Improvement requests.

Steering Committee Review Project and Process Improvement proposals are taken to the Steering Committee and reviewed regularly by business and process owners to determine feasibility, resourcing, and prioritization against other strategic initiatives. The sponsoring business owner and/or Process Improvement Manager assigned to the initiative presents information and recommendations gathered throughout the intake process to the Steering Committee. The Steering Committee then evaluates the request against the other work that is assigned and the backlog within the organization and sequences these priorities optimally. The Committee also reviews the team's available capacity and skills to determine if the work

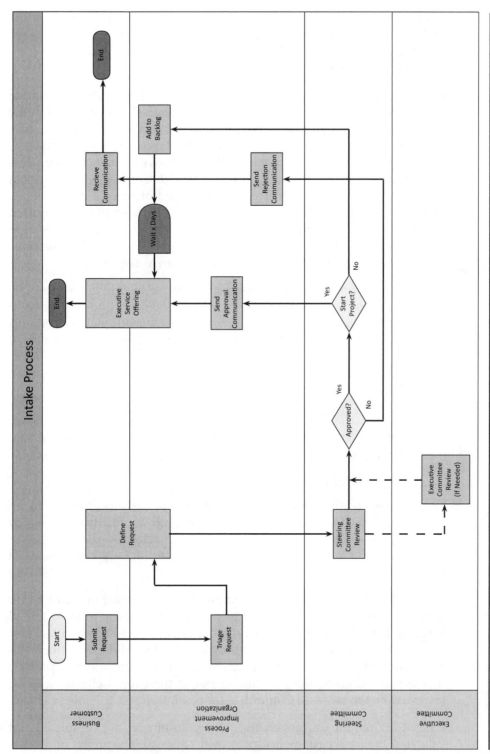

Figure 7-12 Process Improvement Intake Process

193

can be started immediately, if the new request should be placed on the backlog list for future assignment, or if the new request should be cancelled. Proposals are fielded by the Steering Committee and escalated for Executive Committee review.

Executive Committee Review   The Executive Committee serves as the final review stage for major projects or process initiatives. Approval from the executive committee ensures complete alignment to an organization's Strategic Plan and Road Map and signals the start of the Execute phase of the project or process improvement effort. Reviews of project and process performance metrics also occur at the Executive Committee level so that the intended benefits of initiatives are reviewed or reevaluated if unsuccessful. Revisions to the Strategic Plan and Road Map may be required based on business conditions inside or outside the organization as well as performance and key metrics from processes or projects. The Executive Committee makes revisions to the strategic plan as required.

Execute   The execution of the project or initiative begins upon executive committee endorsement. Depending on the scope and effort required to complete the initiative, this effort follows one of the methodologies or techniques outlined throughout *The Process Improvement Handbook*. These include

- Just Do It
- Kaizen
- Lean Six Sigma
- Rummler–Brache
- Root Cause Analysis
- Process Mapping

The purpose of the execution phase is to successfully implement all improvements. It is also to ensure that all individuals, teams, and/or departments across the business that are impacted by the change are trained and made aware of the process and technology changes made during execution.

Monitor   Upon completion of the Execute phase, the warranty and monitoring phase begins. The warranty period describes a predetermined length of time during which performance after project implementation is measured against the expected target. Measurement and evaluation techniques are selected during the project execution stage, and any projects that fail to recognize the benefits set out in the project charter may be reviewed by the Steering Committee for resolution.

Closeout   When the process changes or enhancements are completed and validated, business customers signify their approval. The project or process improvement request is then moved to a closed queue that tracks these requests for historical purposes. The purpose of the Project Closeout phase is to complete the project and finish any final project tasks. This is done in a way that

- Allows for learning opportunities for future projects
- Ensures that project deliverables met business expectations and requirements
- Ensures that business value was realized as expected

This stage also includes the transition of monitoring KPMs to the business unit. Business Owners and Business Leaders are trained to notice irregularities in process performance, project performance, business unit performance, and/or overall organizational performance. These irregularities are identified through KPMs and dashboards within the Process Ecosystem and can result in a review of the strategic plan and subsequent improvement efforts.

Performance Monitoring

Key Performance Measures   A set of measurements must be established at each level in the Governance Model. High-level strategic measurements ensure the organization is delivering value to customers and shareholders, while lower-level measurements ensure project or process initiatives are delivering improvements as intended. High-level strategic performance measures for the organization may include customer satisfaction, on-time delivery, or perfect order fulfillment. These high-level measures are then subdivided into measurements specific to Process Performance, Project Performance, and Business Unit Performance.

Process Performance   The monitoring of processes and procedures within the Process Ecosystem provides indicators as to whether an organization's strategic goals are on track or need review. Process Performance is measured using two tools, Dashboards and KPMs. These tools provide at-a-glance views relevant to processes and procedures. They illustrate summaries, key trends, comparisons, and exceptions. By monitoring weekly cadence and using real-time dashboards, out-of-control processes can be identified quickly. Dashboards support Business Owners and Business Leaders at any level in the organization and provide a quick overview of the organization's health and opportunities.

Project Performance   Similar to Process Performance, Project Performance monitors the projects within an organization. Through a series of Dashboards and KPMs, an organization and its leaders are able to monitor a project's overall health and identify any major or minor roadblocks. This information rolls up to the Business Owners and sponsors to alert the

Steering Committee to any current state risks to the business and signal whether any strategic goals need to be reviewed.

Business Unit Performance   Business Unit Performance measures come directly from the Business Owners and provide an overall health check of the business unit or area. These measures identify any roadblocks within that section, which may come from the process or project performance living within. Any irregularities are reported to the Steering Committee and Executive Steering Committee to review strategic plans and determine if additional goals need to be set.

Enterprise Performance   This is the organization's performance at its highest level. This overarching monitor oversees the Supply Chain strategies for any blocks and routinely monitors identified watch spots. Process, Project, and Business Unit Performance flow up to Supply Chain Performance and provide an overall health check of the organization and its strategic goals. With the established close communication between business units and the levels of governance, improvement efforts can be implemented at the first sign of process disconnect. It is the role of the Business Owners and Business Leaders to inform leadership when processes/procedures/systems go off track and to propose the current state risks.

Governance Roles and Responsibilities

As a Project Request or Change Request moves through the governance process, clarity regarding accountabilities and responsibilities is integral to ensuring a seamless business customer experience. The following section outlines a set of archetypes that describe the roles and functions involved in making process improvement decisions.

Executive Committee   The Executive Committee performs the following functions:

- Ensures alignment between all organizational levels
- Defines strategy and direction as well as measures and targets for the organization
- Reviews the organization's performance and compares it to strategic plans in order to determine if strategies need to be revised or additional plans set
- Sequences and sets priorities as well as resource projects with the necessary staff
- Defines desired business outcomes, strategies, and corporate road maps
- Monitors progress, stakeholders' commitment, results achieved, and organizational performance

- Acts to steer improvement efforts, removes obstacles, manages critical success factors, and remediates process or benefit-realization shortfalls
- Ensures that decisions support normal business operations and blend with organizational culture

Steering Committee   The Steering Committee performs the following functions:

- Evaluates improvement proposals in order to select those that are the best investment of funds and resources
- Sequences and sets priorities as well as resource projects with the necessary staff
- Provides direction on scope and budget of projects
- Monitors progress, stakeholders' commitment, results achieved, and leading indicators of failure
- Acts to steer improvement efforts, removes obstacles, manages critical success factors, and remediates process or benefit-realization shortfalls
- Ensures that decisions support normal business operations and blend with organizational culture
- Evaluates intake request proposals in order to select those that are the best investment of funds and resources and are within the organization's capability and capacity to deliver
- Defines the "desired business outcomes" (end states), benefits and value, business measures of success, and the overall value proposition
- Ensures that decisions support normal business operations and blend with organizational culture

Business Customers and Stakeholders   The Business Customers and Stakeholders perform the following functions:

- Submit project opportunities and improvement ideas
- Align business unit goals and objectives with new opportunities and improvement requests
- Participate in process/training needs analysis and design
- Support in setting project priorities
- Provide any additional process or improvement information as required

Process Improvement Organization   The Process Improvement Organization performs the following functions:

- Evaluates Process Improvement proposals in order to recommend to the Steering Committee those that are the best investment of funds and resources and are within the team's capability and capacity to deliver

- Proposes project priorities to the Steering Committee for approval and resource projects with the necessary staff

- Summarizes and recommends the project schedule to the Steering Committee for approval

- Ensures the Steering Committee is briefed about project proposals, project statuses, and issues in a timely manner

- Chair the Steering Committee

- Confirms validity of process and improvement requests

- Determines complexity/estimated duration

- Ensures requests align with organizational and strategic goals

- Drives conversations with Business Customers and Stakeholders to ensure accurate scope

- Establishes the basis for process governance, approval, and measurement, including defining roles and accountabilities, policies and standards, and associated processes

- Leads the management review to validate process opportunities and improvement requests

- Leads prioritization efforts

Implementing Governance

Every enterprise is unique and should determine the best way to implement its own governance structure. Following are suggested steps for organizations to consider as they implement governance.

1. *Determine principles and goals:* An organization should initially develop a governance vision, governance policies, as well as standards that can be measured to track compliance and quantify the benefit of governance activities to the enterprise. For example, at this stage, organizations might outline the individuals who should be included as attendees of any steering committee meetings, outline what the governance process might entail, and publish and publicize the principles, goals, and standards.

2. *Develop an education strategy:* The governance policies and procedures that are established should be accompanied by ongoing education and training plans. Note that this includes training in the use of any project management tools and templates as well as training in the governance standards and practices. For example, at this stage, organizations might create Frequently Asked Questions Pages, roll out formal training workshops, or publish online

tutorials that describe the implementation and use of the organization's governance system.

3. *Develop an ongoing plan:* Because successful governance is ongoing, the governance body should meet regularly. Ongoing activities include incorporating new requirements into the governance plan or reevaluating and adjusting governing principles and standards. Conflicts may need to be resolved as competing needs arise between various business divisions. The governance steering committee should also report regularly to the executive committee; this will help foster accountability and assist with enforcing compliance across the organization. The goal of governance is to increase the return on investment in governance, maximize the usefulness of the organization's workforce, and increase the organization's overall performance.

Department Controls

Internal control involves the integration of a department's activities, attitudes, plans, policies, and efforts in order to provide reasonable assurance that the department will achieve its mission. Simply put, internal control is what a department does to ensure that the things they wish to happen do happen and the things they don't wish to happen don't happen. Controls within a Process Improvement organization exist to ensure that organizational practices are consistently reviewed and managed and that any internal processes, templates, or artifacts are maintained over time. These controls help foster a culture that is process driven and focused on meeting customer requirements and creating predictable results. Enabling controls within a PIO ensures that

- All processes and procedures are documented and regularly reviewed

- Feasible and timely response plans are created in the event that a problem with a process arises

- Replication opportunities are reviewed and any duplication of effort is removed

- Process quality is linked directly to customer requirements and strategic objectives rather than functional success

- All necessary stakeholders are included in the decision-making process

- Orderly, economical, efficient, and effective operations are promoted

- Quality products and services that are consistent with the department's mission are produced

- Resources are safeguarded against loss due to waste, abuse, mismanagement, errors, and fraud
- Reliable financial and management data are developed and maintained and reported in a timely manner
- Successes are celebrated

Typical PIO Controls are described in the following paragraphs.

Process documentation: A typical improvement control that is regarded as the most critical component of successful PIOs is to ensure that standard practices and processes within the PIO are documented. Documentation is critical because it

- Describes the flow of processes and standardizes the procedures for operating the process
- Provides approved information for all current and future employees who will need it
- Increases organizational learning and provides training materials
- Is a necessary step to ensure that the knowledge gained via process improvement projects is shared and institutionalized

Centralized process tracking: Any process that spans across an organization requires orchestration and visibility. Additionally, any process that has direct impact on revenue, customer experience, or how employees conduct their day-to-day activities should be tracked in a centrally accessible location.

Often, organizations have several, if not hundreds, of active and in-life processes that have been documented. Centralized tracking ensures that these processes are monitored, controlled, and easily prioritized if improvements are needed. Centralized tracking can take many forms; the most common tools used are comprehensive software programs such as Process Central, a program developed by iGrafx, or LiveLink, a document repository program developed by OpenText.

Benefits of centralized project tracking include

- *Transparency:* A central process library serves as a repository for all process information, allowing workers and managers to store, manage, and distribute all relevant processes in a consistent and easy-to-use format.
- *Consistency:* Central tracking offers a consistent presentation format that allows process owners, users, sponsors, and other stakeholders to instantly locate needed information.
- *Flexibility:* Tracking systems are customizable and designed to meet an organization's unique tracking needs.

- *Manageability:* Tracking systems allow managers, executives, and steering committee members to see every process in real time, allowing them to shift workloads and redelegate responsibility as needed.

- *Organization:* Tracking systems allow processes to be catalogued into the proper category and to be sorted using any number of fields for targeted discussions and planning. They often include text-searching features as well.

- *Security:* Processes are widely accessible but secure, contain accurate document history, and are easily rolled back to previous versions.

- *Link management:* Document links are automatically maintained and subprocesses are easily retrieved from master process documents.

- *Process sharing:* Process sharing makes it possible to reuse process elements across multiple maps and speed-up process development.

Asset security: Security may be physical or electronic or both. Equipment, artifacts, and other organizational assets should be secured and periodically checked.

Organizational structure: Lines of authority and responsibility should be clearly defined so that employees know who to report to regarding performance of duties, problems, and questions related to their position and the organization as a whole. An organization chart can be used to define this structure as long as it is kept up to date. Part of the structure is also the rules that employees must abide by. Rules provide guidance to employees as they carry out their duties, clear rules on allowable and expected performance, and a means for enforcement.

Authorization and approval: Artifacts should be authorized and approved in order to help ensure activities are consistent with departmental or institutional goals and standards. The important thing to remember is that the person who approves transactions must have the authority to do so and the necessary knowledge to make informed decisions.

Writing styles: Process Improvement artifacts should be written in a concise, step-by-step, easy-to-read format. The information presented should be unambiguous and not overly complicated. The active voice and present verb tense should be used. The term *you* should not be used but implied. Any documentation should not be wordy, redundant, or overly lengthy. Information should be conveyed clearly and explicitly to remove any doubt as to what is required, expected, or being asked.

Review and approval: Process Improvement artifacts should be reviewed by one or more individuals with appropriate subject matter training and experience. Someone at the appropriate level of management should approve the finalized documents when necessary. E-mail or signature approval indicates that the Process Improvement artifact in question has been both reviewed and approved by management.

Revisions and reviews: Process Improvement artifacts need to remain current to be useful. Therefore, whenever processes are changed, associated artifacts should be updated and reapproved. If desired, only the pertinent section of an artifact can be modified and the change date/revision number indicated in the document control notation section.

Process artifacts should also be systematically reviewed on a periodic basis to ensure that the documentation remains current and appropriate or to determine whether certain artifacts are needed. The review process should not be overly cumbersome to encourage timely review. Management should indicate the frequency of review.

Checklists: Many Process Improvement activities can benefit from the use of checklists to ensure that necessary steps are followed. Checklists are also used to document completed actions. Any checklists or forms included as part of an activity should be referenced at the points in the procedure where they are to be used. In some cases, detailed checklists are prepared specifically for a given activity.

Document control: PIOs should develop a numbering system in order to systematically identify and label process documents and artifacts. Generally, each artifact should have control documentation notation. A short title and identification number can serve as a reference designation. The revision number and date are very useful when reviewing historical data and are critical when evidentiary records are needed.

Chapter Summary

In this chapter we presented an outline of the PIO. The chapter described the various services offered by a PIO and the organization's roles and responsibilities. Process Improvement Governance was also discussed, and a formal structure and model was presented that helps ensure both strategic and tactical alignment for all Process Improvement endeavors. From this chapter, we also learned that

- A PIO typically focuses on the following five areas:
  - Enterprise Understanding
  - Workforce Engagement
  - Project Management
  - Process Improvement
  - Continuous Assessment
- A PIO provides a number of professional services to assist organizations achieve process excellence. These services can range from directly managing Process Improvement Projects, to implementing and modeling business process architectures, to simply providing Process Improvement support to business customers by way of guidance and education.

- Most PIOs include the following three roles:
  - Process Improvement Coordinator
  - Process Improvement Manager
  - Process Improvement Architect
- The term *Governance* refers to the set of policies, roles, responsibilities, and processes that an organization establishes in order to direct, guide, and control how the organization uses its workforce, processes, and technologies to accomplish business goals.
- The purpose of Governance is to provide a decision-making framework that is logical, robust, and repeatable in order to govern an organization's projects, processes, resources, and investments.
- An effective Governance structure is a key component of any process-focused organization and enables enterprise-wide process improvement and transformation.
- The Process Improvement Governance is comprised of the following seven parts:
  - Forming an executive committee
  - Creating a strategic road map
  - Building management steering committees
  - Managing the Process Improvement Project Portfolio
  - Managing the Process Backbone and Ecosystem
  - Executing the organizational governance process
  - Monitoring process and organizational performance
- The monitoring of processes and procedures within the Process Ecosystem provides indicators as to whether an organization's strategic goals are on track or need review.
- As a Project Request or Change Request moves through the governance process, clarity regarding accountabilities and responsibilities is integral to ensuring a seamless business customer experience.
- Internal control is the integration of a department's activities, attitudes, plans, policies, and efforts in order to provide reasonable assurance that it will achieve its mission.

Chapter Preview

Now that the components of the PIO have been discussed, the next step is to learn the various skills, competencies, and tools demonstrated most often by Process Improvement professionals. Chapter 8 provides a comprehensive dictionary of aptitudes exhibited in Process-focused organizations.

CHAPTER 8

Process Improvement Aptitudes

This chapter provides a comprehensive lexicon of skills, competencies, and techniques that managers and practitioners can draw upon as they execute Process Improvement efforts. Whether an individual is developing a business architecture, creating a process model, leading a Process Improvement project, or simply participating in Process Improvement efforts, possessing the necessary tools, skills, and competencies to effectively contribute and provide value to these activities is paramount. Process Improvement professionals need to keep participants engaged, elicit necessary information, create shared understanding, and build consensus, all while keeping stakeholders focused on the intended goal of their engagement. This chapter identifies the skills, underlying competencies, and the techniques needed to drive results within a process-oriented enterprise and provides an understanding of the factors that comprise a competent process-focused individual. This chapter is organized around the following six major topics:

- *Overview of the Dictionary:* What is the Process Improvement Aptitudes Dictionary? How does it relate to Process Improvement efforts?

- *Skills and Competencies Defined:* What is the definition of a skill? How is a skill different from a competency?

- *Process Improvement Skills:* What skills should Process Improvement professionals possess?

- *Process Improvement Competencies:* What competencies are required in order to effectively execute Process Improvement initiatives?

- *Process Improvement Tools and Techniques:* What are the most common tools and techniques used in Process Improvement efforts?

- *Building, Recognizing, and Retaining Talent:* What key practices can be used to help build and sustain employee engagement and retention?

Overview of the Dictionary

Organizations that are focused on improving their processes and operations should make a conscious effort to properly identify and conceptualize the skills, competencies, and techniques required by their workforce in order to ensure widespread cultural practice. The characteristics that are often displayed by successful Process Improvement professionals are not simply personality traits or skills that have been learned over time. These characteristics are, instead, patterns of behavior and are a combination of skills, knowledge, tools, and personal attributes. The *Process Improvement Aptitudes Dictionary* strives to outline the underlying characteristics that define the patterns of behavior required for Process Improvement professionals to deliver superior performance. There are ten skills and ten competencies that are considered essential for ensuring the success of individuals involved in Process Improvement efforts. When combined with various Process Improvement methods and techniques, these skills and competencies can be used collectively to deliver process excellence.

The skills, competencies, and techniques outlined in this chapter serve to provide a

- Common language of behaviors, skills, knowledge, and techniques that process-focused organizations can use to ensure performance success

- System for identifying superior workforce performance within a process-oriented organization

- Set of measurable performance criteria that any employee involved in Process Improvement activities can demonstrate

Likewise, this dictionary is intended to serve as an anchor for Process Improvement capabilities and provides a common reference point for individuals who are directly involved with Process Improvement efforts.

Using the Process Improvement Aptitudes Dictionary

Each component of the dictionary contains the following 2 key elements:

- *Overview and Definition:* Each item includes an overall definition that explains, in general, the characteristics, traits, or motives associated with the skill, competency, or technique.

- *Examples:* Each component is accompanied by a set of generic examples that describe what the skill, competency, or technique might entail or how it is used.

While this chapter focuses on a specific set of aptitudes needed by Process Improvement practitioners and how to recognize and build these aptitudes within an organization, there may be other factors not identified that are uniquely required in various professional scenarios and environments. The Process Improvement Aptitudes Dictionary offers a baseline for organizational application and is presented as a sample set of attributes that reflect what outstanding Process Improvement professionals demonstrate most often.

To obtain the full benefits of the aptitudes outlined in this dictionary, readers should

- Recognize and understand the aptitudes:
  ○ Review the definitions of each skill, competency, and technique
  ○ Learn to recognize each aptitude as it is displayed in other individuals
- Practice the aptitude:
  ○ Continuously use the skills, competencies, and techniques in order to improve and enhance their effectiveness

Skills, Competencies, and Techniques Defined

An important consideration when reading or using the aptitudes found in this chapter are the definition of the terms, skills, competencies, and techniques and how they relate to each other. One distinction between a skill and a competency is that a competency is more than just knowledge or skills. It involves the ability to meet complex demands by drawing on and mobilizing psychosocial resources, such as skills and attitudes, in a particular setting. In essence, a competency refers to a set of skills, and is more of an umbrella term that also includes behaviors and knowledge, whereas skills are specific learned activities that may be part of a broader context. For example, the ability to be flexible in business situations is a competence that may draw on an individual's knowledge of change management principles, techniques, and practical skills and attitudes. A skill can be thought of as an individual's ability to perform tasks or solve problems, while techniques are the various ways of carrying out a particular task. For example, Process Improvement practitioners may require strong facilitation skills, yet conduct facilitation activities using a particular approach or technique, such as Brainstorming. A competency is therefore a broader concept that, in many cases, is comprised of skills as well as attitudes, knowledge, and techniques. All three areas are very much aligned and serve common purposes. The purpose of this dictionary is to provide as broad an area as possible in defining the various aptitudes demonstrated or used by Process Improvement professionals.

Process Improvement Skills

The skills needed to objectively assess and improve processes range from basic management to advanced technical and project management expertise. The following section outlines the skills that seasoned Process Improvement practitioners, managers, and operators typically possess.

Facilitation Skills

Facilitation describes the process of taking a group through learning or change in a way that encourages all group members to participate. This approach assumes that each person has something unique and valuable to share. Without each person's contribution and knowledge, the group's ability to understand or respond to a situation may be reduced. Facilitation skills are important to Process Improvement efforts as improvements are typically executed in a group setting either through large meetings or workshops. In these scenarios, a facilitator's role is to draw out knowledge and ideas from different group members, encourage group members to learn from each other, and encourage group members to think and act together as they propose improvements to organizational processes and systems.

An individual who possesses facilitation skills

- Uses group discussion and activities that allow stakeholders to be actively involved in a particular task or initiative
- Prepares for group meetings by identifying the key issues, goals, and stakeholder expectations
- Identifies resources that are most likely to help a group with its task, clarifies the agenda and objectives, and allocates the necessary time to cover topics
- Recognizes the strengths and abilities of individual group members and helps them to feel comfortable about sharing their hopes, concerns, and ideas
- Supports the group, giving participants confidence in sharing and trying out new ideas
- Engages all members in the discussion and builds on the ideas of contributors, while ensuring other members are not overwhelmed or discouraged from giving input
- Sees when the group is off track and redirects the conversation toward productive channels
- Values diversity and is sensitive to the different needs and interests of group members

- Allows ownership of the process by group members, highlights group successes, builds a sense of shared accomplishment, and reinforces success by becoming an advocate for the group's decisions
- Leads by example through attitudes, approach, and actions

Change Management Skills

Change Management is the process of transitioning individuals, teams, and organizations from a current state mentality or way of operating to a desired future state. This approach serves to alter human capability or organizational systems in order to achieve a higher degree of output or performance. By not properly managing change, organizations risk incurring project delays, employee resistance, decreased morale, and increased expenses. A change agent's role is to enable people to do more, find new and better ways of operating, and manage the activities surrounding any change within an organization's environment in order to avoid stakeholder frustration and fatigue.

An individual who possesses change management skills

- Acts as a catalyst for change on a consistent basis
- Defines, researches, plans, builds support, and partners with others to create change
- Has a clear vision for the future and is able to communicate that vision clearly
- Regularly assesses current state or common practices and suggests possible change initiatives
- Builds bridges of trust and forms strong relationships with stakeholders
- Possesses courage and a willingness to do what is best for the organization
- Acts as an advocate and promoter of change whenever possible
- Supports and encourages transformation activities

Project Management Skills

Project Management describes the process of planning, organizing, motivating, and controlling resources and activities in order to achieve specific goals. This approach involves defining the scope of a particular endeavor, building a schedule for executing its activities, managing the effort in a disciplined manner, and ensuring appropriate controls and quality measures are in place. By not properly managing project activities, organizations risk duplicating efforts, causing confusion among stakeholders, and increasing expenses. A critical component of being a Process Improvement Manager is to act as the primary point of contact for a project and to formally manage the timeline, costs, and risks associated with its execution so that project deliverables meet or exceed stakeholder expectations.

An individual who possesses project management skills

- Aligns activities with business goals
- Integrates the ideas and needs of others when developing feasible strategies to achieve goals
- Ensures the goals, purpose, and criteria for success are clearly defined
- Ensures needed resources and skill sets among staff are available
- Removes obstacles and motivates team members
- Outlines clear expectations, roles, and responsibilities for initiatives
- Identifies potential risks and develops plans to manage or minimize them
- Keeps stakeholders informed and up-to-date with regular meetings and reports
- Manages costs and budgets
- Manages schedules and ensures deadlines are met

Time Management Skills

Time Management is the process of planning and exercising control over the amount of time spent on specific activities in order to increase effectiveness, efficiency, or productivity. This involves prioritizing and selecting what to work on and when to work on it, as well as ensuring that tasks are completed within appropriate timelines. By not effectively managing time, individuals risk wasting effort on low-priority tasks and deliverables, becoming distracted by items of little or no importance, and completing tasks behind schedule.

An individual who possesses time management skills

- Sets aside time for planning and scheduling of activities
- Uses goal-setting to decide what tasks and activities should be worked on
- Groups related tasks to be more efficient
- Adjusts priorities as situations change
- Easily transitions between tasks and picks up where left off when interrupted
- Evaluates progress on tasks and adjusts work style as needed
- Completes high volumes of work, keeping a rapid pace without sacrificing accuracy

- Leaves contingency time in timelines to deal with unexpected issues or events
- Regularly confirms priorities with leadership and stakeholders
- Ensures the value of tasks outweighs the effort needed to deliver them

Analytical Skills

Analytical Thinking is a process that emphasizes breaking down complex problems into single and manageable components. It involves gathering relevant information, identifying key issues related to this information, comparing sets of data from different sources in order to identify possible cause-and-effect patterns, and drawing appropriate conclusions from those datasets in order to arrive at appropriate solutions to problems. Problem solving based on data and fact-finding techniques is an essential competency to Process Improvement efforts. Too often, cursory examinations of situations result in glossing over important facts or misleading the assessor based on personal experience and bias. Approaching process work with agnostic and impartial information-gathering establishes a data stream that can be monitored for trends and patterns that will help in identifying root causes.

An individual who possesses analytical thinking skills

- Takes care to define each problem carefully before trying to solve it
- Sees relationships between information in varied forms and from varied sources
- Notices when data appear wrong or incomplete and isolates information that is not pertinent to a decision or solution
- Breaks down complex information into component parts, sorts and groups data, and applies causal relationships
- Strives to look at problems from different perspectives and generates multiple solutions
- Tries to address the political issues and other consequences of a change so that others will understand and support the solution
- Evaluates potential solutions carefully and thoroughly against predefined standards
- Systematically searches for issues that may become problems in the future

Negotiating Skills

Negotiation is a process that encompasses a discussion between two or more individuals who seek to find a solution to a problem that meets both of their needs and interests. Being a skilled negotiator can help individuals solve problems, manage conflict, and preserve relationships. In any disagreement, individuals understandably aim to achieve the best possible outcome for their position or organization. However, the principles of fairness, seeking mutual benefit, and maintaining a relationship are key to a successful outcome. Negotiation skills can be of great benefit in resolving any differences that arise between project team members and stakeholders.

An individual who possesses negotiation skills

- Presents interests in ways that foster the understanding and resolution of problems
- Gains the trust of other parties by being honest, respectful, and sensitive to their needs
- Knows when to be gentle and when to be assertive and acts accordingly while avoiding ultimatums
- Questions and counters proposals without damaging relationships
- Remains open to many approaches in order to address needs or resolve issues and seeks suggestions from other parties
- Seeks common interests and win–win solutions

Decision-Making Skills

Decision Making is the process of choosing what to do by considering the possible consequences of different choices. Many different factors can influence how individuals make decisions, including cognitive, psychological, social, cultural, and societal factors. Decision Making involves the ability and willingness to be unbiased and objective when making sound business decisions based upon consideration of various alternatives. This includes consideration of both long- and short-term impacts of decisions on various individuals or groups. Poor Decision Making in Process Improvement efforts can lead to poor-quality results, financial costs, or unnecessary waste.

An individual who possesses decision-making skills

- Determines the actual issue before starting a decision-making process
- Considers lessons learned from experience, differing needs, and the impact of the decision on others
- Takes the time needed to choose the best decision-making tool for each specific decision
- Balances analysis, wisdom, experience, and perspective when making decisions
- Considers a variety of potential solutions before making a decision

- Finds solutions that are acceptable to diverse groups with conflicting interests and needs

- Evaluates the risks associated with each alternative before making a decision

- Determines the factors most important to the decision and uses those factors to evaluate options

- Has the ability to explain the rationale for a decision

- Makes necessary decisions even when information is limited or unclear

Communication Skills

Communication is the process of conveying information through the exchange of speech, thoughts, messages, text, visuals, signals, or behaviors. Clear and informative communication is critical during periods of change. It provides stakeholders with all of the information needed to ease the transition from the current state to the future state. Effective communication takes into consideration the needs of the audience and their concerns and the communication channels best suited to the audience, has a clear strategy that defines the approach, and establishes the infrastructure needed to deliver the communication strategy. Various channels of communication influence audiences in different ways and aid in the reinforcement of the intended core message. Such tools include social websites, corporate intranets, video messages, real-time broadcasts, commercials, posters, e-mail, physical mail, meetings, and so on. A communication plan helps to describe the unique elements of the communication and sets the roles and responsibilities for where, when, how, and who will communicate the message. A solid plan never assumes communications will flow by themselves; it orchestrates all aspects of communication to ensure everything will progress on track. The plan creates the standards for and actual message tone that is used to remind leadership of the program's benefits and the value-add to their organizations. Competent communicators attend to this competency as a critical factor to the overall transformation made possible by process improvement work. These specialists bidirectionally process outbound communication and act as active listening agents to identify information breakdowns and areas requiring adjustment. Through active listening one gains insight, understands what is expected and needed, discovers the opportunity to resolve problems, and influences the ecosystem to become more productive. Attention needs to be given to communication style when interacting with audiences that are globally distributed and diverse. The origination of communication messages should be reviewed carefully to screen for country-specific/audience-specific sensitivities that are often missed in language translation and inflection to avoid offense or misdirected action.

An individual who possesses communication skills

- Thinks about what a person or audience needs to know and how best to convey it
- Communicates concisely and repeatedly using appropriate means (e.g., group/one-on-one, e-mails/meetings, face-to-face)
- Ensures all relevant background information and detail are included so that the message is understood
- Uses diagrams and charts to express ideas as required
- Continuously updates stakeholders as new information becomes available and lessons are learned
- Helps people understand the underlying concepts behind the point being conveyed

Training Skills

Training is a process aimed at bettering the performance of individuals and groups in organizational settings. It is the act of formally delivering information, knowledge, and skills to individuals or groups so that they are able to perform duties appropriately. Lack of training in Process Improvement efforts can lead to confusion, poor performance, and undesired outcomes.

An individual who possesses training skills

- Creates an environment for optimal learning
- Tailors his or her teaching style to the audience in question
- Combines exercises, group discussions, workshops, and other methods to meet diverse learning styles
- Uses props, slides, and other presentation aids to deliver new content and information
- Interacts with the audience, reads body language, gathers feedback, and holds their attention
- Sees when listeners fail to grasp critical concepts and take steps to ensure comprehension
- Gives adequate attention to individuals without neglecting the group as a whole

Coaching Skills

Coaching is the process of enabling stakeholders to grow and succeed by providing feedback, instruction, and encouragement in order to assist them in discovering solutions on their own. A coach helps stakeholders articulate their goals, define strategies and plans, and challenge as required in order to stay the course and succeed.

An individual who possesses coaching skills

- Coaches others regardless of performance level
- Shares specialized approaches and skills that will increase capabilities
- Helps others identify key goals and use their talents to achieve those goals
- Sees the potential and strengths of others and works to build on them
- Actively supports others as they stretch beyond their comfort levels and try new techniques that may enhance success
- Coaches for incremental, one-step-at-a-time improvements, offering praise and recognition as each step is taken
- Suggests methods and gives examples that provide a road map to improved performance
- Models success behaviors, a high-performance work ethic, and constant self-improvement

Process Improvement Competencies

Core *competencies,* also referred to as key competencies, consist of those behaviors and capabilities an individual has that make him or her stand out as a superior performer. Some relevant competencies demonstrated by Process Improvement professionals and stakeholders are described in the following paragraphs.

Building Partnerships

Building partnerships involves the ability to build mutually beneficial business relationships that foster improved business outcomes. Partnerships imply a sense of cooperation rather than competitiveness and strive for balance within the established trust. Effective partnerships take a long-range approach and recognize the nature of give-and-take based on the current needs of the organization. It is not unusual for imbalance to occur from time to time, but overall successful partnerships are maintained because all partners receive a significant benefit from the relationship. Process work requires driving for deep and lasting partnerships between colleagues, department, suppliers, and customers. This is achieved by demonstrating comprehensive knowledge of the business strategy, sources of revenue, and methods/processes to perform the service.

Competency examples include

- Responding to the needs of customers and stakeholders in a timely manner
- Ensuring customers and stakeholders know how to access help when required

- Acting as a partner and trusted advisor
- Encouraging others to involve customers in process development
- Building reciprocal relationships with people in other functions
- Sharing experience and expertise with contacts in other organizations
- Working with external colleagues to foster the transfer of process information, which results in business improvement in both directions
- Creating and leading cross-organizational teams to look at new business opportunities

Collaboration

Collaboration embraces the ability to work harmoniously with others in the business environment toward a common goal. As in partnership relationships, collaboration requires a cooperative spirit of participation and ease of doing business. It leverages appropriate information-sharing and knowledge transfers across functional boundaries to effect positive results. Collaborators build necessary relationships proactively and reactively. They are self-aware of their personal needs and set them aside, and they expect those they collaborate with to do the same. Their energy level is heightened when teaming with others to solve a business problem. Collaborators seek information for the purpose of understanding, and they offer insight freely to help shape an outcome that is not recognized as a personal win. They radiate information and teach others in a coaching-like manner. Those with strong collaboration skills understand their own role in defusing conflict and helping others to align in order to meet the needs of the customer or other stakeholders and achieve business success. The expert collaborator probes for improvement while not creating blame or excessive discomfort in the status quo but effectively moves the business forward. There is a personal likability associated with most successful collaborators as they bridge opposing perspectives to get results while creating the sense of a winning team around them.

Competency examples include

- Disclosing pertinent information freely to keep colleagues abreast of activities
- Contributing to team meetings and discussions
- Maintaining positive working relationships with others by showing interest in what others are doing
- Providing opportunities for others to present the results of their own accomplishments
- Understanding what motivates different colleagues and using that knowledge to select the most effective recognition vehicle for them

- Intervening when conflict arises to help those involved bring up their issues in order to get to the root causes of the problem
- Supporting joint projects or sharing of resources
- Seeking out opportunities for cross-functional collaboration
- Promoting open discussion by inviting and sharing ideas and responding to others with empathy

Credibility

Having organizational *credibility* means having the ability to establish the trust needed to help organizations move through the transformation stages more smoothly. This is built on a history of delivering service with integrity, competence at execution, the right relationships within and outside of the organization, as well as career experience or educational status. Credibility can be eroded if the individual put in the position of responsibility acts in a contrary manner, that is, is insincere, unethical, secretive, mean spirited, and so on. When this occurs, the worker's integrity and their body of work is subject to suspicion. The credible worker acts as a role model and inspiration for others, which motivates others to action. Their work is considered best in quality and drives others to follow in their footsteps. Having credible processes in place helps maintain the organization's credibility by delivering processes that are consistent and compliant with the rules and laws and that do not intentionally disadvantage others. Inspecting for attempts of exploitation or gaps in processes helps to test for credibility issues and reinforces the expectation for credible conduct.

Competency examples include

- Consistently delivering what is promised
- Consistently abiding by company standards, policies, and procedures
- Listening actively and openly to the views and opinions of others
- Constructively presenting ideas even if they are different from accepted opinion
- Acting as a role model for others through moral and ethical integrity
- Encouraging and recognizing others who behave ethically
- Focusing on solving a problem and not assigning blame

Flexibility

Flexibility describes the ability to adapt to change, shift focus and resources, and manage through the change. It is important to have flexibility as change is constant in the competitive business environment. Individuals who are flexible keep an open mind to alternatives so that they can create an environment in which innovation thrives and problems can be solved without stress. They seek out feedback to understand and appreciate different and opposing perspectives on an issue and adapt their approach accordingly. Processes should be built to recognize the nature of change within long-term strategy and market and technological trends in order to capitalize on potential opportunities. As implied, flexibility is correlated with speed and agility. The ability to shift on demand in response to unanticipated market changes and competitive pressures is essential. While not all outcomes can be anticipated in a volatile market or environment, building flexibility skills in the workforce creates the expectation for speed and flexibility in service delivery.

Competency examples include

- Responding positively to change and embracing and using new practices or values to accomplish goals and solve problems

- Adapting approaches, goals, and methods to achieve solutions and results in dynamic situations

- Coping well and helping others deal with the ongoing demands of change

- Managing change in a way that reduces the concern experienced by others

- Remaining optimistic and positive when experiencing change

- Continuing to deliver expected results when priorities and resources are redefined

- Cooperating when activities, plans, and timelines need to be reprioritized

- Reassuring others when change occurs by sharing information openly and helping others to understand the reason for change

- Conducting team-building sessions in order to help the team adapt and remain positive when change occurs

- Consulting with team members for thoughts and ideas to a situation before implementing a change

Initiative

Initiative is the ability to take immediate action regarding a challenge, obstacle, or opportunity while thinking ahead to address future challenges or opportunities. People with initiative identify trouble spots and take steps to disrupt the eventual result of the problem. They are driven to prevent problems where possible and to take advantage of various tools at their disposal to do so. This is done with great personal investment in the outcome. Organizations that operate suggestion boxes or programs for channeling innovative ideas often are surprised at the high level of engagement among some of their staff who generate comprehensive and thoughtful recommendations about problems they experience in their jobs. These employees are demonstrating a high bias for action and are at the forefront of employees with positive initiative. Process work requires a great deal of initiative to drive the business to take immediate action that will result in long-term benefit, even in the absence of short-term gains. In this context it is important to recognize the momentum of a company, which is behind maintaining the business in an as-is state. Initiative is a key driver that gets others to pay attention to the to-be state and the advantages of making the change.

Competency examples include

- Taking repeated actions to achieve goals despite facing obstacles and not giving up easily

- Demonstrating a high degree of perseverance to ensure goals are reached and at the required standard

- Finding another way around obstacles when blocked

- Recognizing opportunity and being able to act accordingly to translate ideas into a functional reality

- Acting proactively before being forced by a crisis

- Actively seeking solutions before being asked or directed

- Acting quickly to address current deficiencies in the organization and creating an awareness that spurs others to respond

- Questioning the way things are done and taking action to lead toward increased performance

- Creating a sense of urgency in others and ensuring actions are taken through appropriate influencing and rationalization strategies

- Acting on changes, trends, and emerging issues that will impact the organizational process

Business Knowledge

Business Knowledge means having the business acumen needed to drive strategy, evaluate, and improve performance. Those who don't possess a solid understanding of the business they are operating within are at a tremendous disadvantage if left in that state. Using the process landscape is a great method for gaining knowledge about the business. The ecosystem that sustains the business, the inputs and expected outputs, the performance information that the system radiates, and the feedback mechanisms contained within are all opportunities to first learn the basics and drive attention to the elements of the business that are often neglected. Processes often start out being fairly mechanical; over time, significant understanding of how a business is constructed is gained. Initially people with whom the process team interacts see a naive team that records information with fervor. However, as business knowledge grows, the business team is challenged to keep pace with the process organization. As the knowledge depth and breadth expand, the process organization probes deeply into the business as-is state to reveal significant gaps in the understanding of why the business operates the way it does. These are the moments when revelations arise and business members are surprised to have the process team draw their attention to very real opportunities to run their business more effectively. The process team and the artifacts they produce become a hub of business intelligence that tenured staff depends on and embraces. Those who are new to the business landscape learn their place in the value stream more quickly.

Competency examples include

- Obtaining general knowledge in all areas of the business
- Applying cross-functional knowledge to solve concrete business problems
- Describing the responsibilities of management in relation to the organization's goals and strategy
- Understanding the market customer base and environmental factors of the organization
- Outlining the various business goals and how they can be obtained

Sound Judgment

Sound Judgment is the ability and willingness to be unbiased and objective when making business decisions based upon consideration of various alternatives. A critical skill, judgment is expected in almost every job. Quality of judgment is dependent on the quality of information (the facts), accurate analysis and reasoning skills (the assessment), and sufficient time to evaluate all potential outcomes and risks (the options) in order to make a good decision (the decision). Judgment is ultimately measured as a matter of the outcome result. Sound judgment can be tested. For example, mock scenarios can be presented during situational interviews to observe the candidates' methods for making a decision. Information gathering through additional inquiries and consultation, reference to research or experience regarding the scenario, compilation of theories and expected results, evaluation of best-case eventualities, and indications of unbiased action may offer evidence that the candidate would, in practice, have reasonable judgment. The approach one takes does have a qualitative impact on the result. Those who are able to discard unimportant information and distill important information quickly are more likely to achieve business success. Breaking down problems into smaller segments makes it easier to assess the problem more accurately. The cause-and-effect relationship can be more easily identified, pointing to the root cause. Recognition of patterns and trends helps build predictability about future events. Judgment can be improved with experience, quality feedback, and reflection on what could have been improved. This insight can be used to prepare for the next experience.

Competency examples include

- Making sound decisions by considering alternatives
- Considering impact in other areas of the organization
- Weighing alternatives and selecting practical solutions
- Reviewing a decision to see if it satisfies long-range plans

Resilience

Resilience is the ability to remain strong and persistent when pursuing goals despite obstacles and setbacks. It also means accepting the dynamic nature of business and maintaining a positive demeanor while facing the challenges of unknown or difficult situations. A resilient person is able to demonstrate that they are able to reorient themselves and others to deal with a situation. Often thought of as the ability to "bounce back" in the face of adversity, resilience manifests itself in constructive action that embraces lessons learned and sees the opportunity for improvement. With a focus on who is in control, process improvement professionals focus their attention on supporting others in the rapidly changing environment. They encourage new ideas, perspectives, strategies, and positions and take steps to understand the reasons for changes in the environment. Being resilient requires a level of creativity and willingness to reconsider strongly held ideas and beliefs.

Competency examples include

- Expressing realistic optimism to others
- Remaining positive and upbeat during setbacks
- Not showing undue frustration when resisted or blocked by others or by circumstances
- Putting obstacles into perspective
- Remaining composed under pressure
- Reflecting on setbacks and identifying key lessons learned for self and others
- Encouraging others to see more hopeful perspectives and outcomes
- Coping well with ambiguity and knowing how much to push people and situations and when to let go
- Keeping individuals focused on what they can control and on the improvements they can make

Strategic Implementation

Strategic Implementation is the ability to link strategic concepts to daily work. Understanding the organization's Strengths, Weaknesses, Opportunities, and Threats (SWOT), a strategic thinker balances SWOT and industry/ market trends to turn the business toward long-term sustainable competitiveness. They construct processes that are congruent with the corporate strategic plan and leverage the process landscape as a mechanism to both direct the workforce in achieving results aligned with the plan and use the landscape to monitor and alert others regarding performance trends. The information that radiates from the performance measures helps shape strategic change to business plans as they help identify changes in the competitive landscape.

The priority of process work is to take into account the immediate actions and tactics that are driven by the business priorities and enable others to understand the priorities and enlist support for implementing new approaches to meet these needs. The Process-Oriented Architecture model is a powerful instrument that is used in constructing transition architectures that align with long-term strategy results. It captures the essential components of what the future vision is believed to be when fully operational. In this way, current state, transition architecture models, and to-be state work in concert to support the evolving strategy of the business.

Competency examples include

- Focusing on immediate tactical action to achieve near-term results while keeping long-term objectives in mind
- Prioritizing work in order to meet expectations
- Adapting his or her role, when appropriate, to the organizational business plan
- Working with customers and stakeholders to understand their business strategies
- Anticipating future situations to ensure required processes are in place
- Establishing a realistic set of actions and milestones to ensure success
- Finding the needed resources to support implementation
- Anticipating future trends and consequences accurately
- Regarded as a thought leader
- Describing a clear vision for the future and a compelling picture of new opportunities

Situational Awareness

Situational Awareness is the ability to make sensible decisions based on an accurate understanding of the current environment. Situational awareness leverages human characteristics such as perception, which can vary greatly from person to person. It can be interfered with emotional state, historical experience, fatigue, pressure driven from social norms, bias, expectations, and/or stresses interrupt the process of being able to make an accurate assessment of the current situation. Possessing good situational awareness may not be consistent from occasion to occasion, as even the best practiced individuals can suffer one or more of the influences that interfere with understanding. Time available is also a consideration; making a snap decision inherently reduces the variables that can be considered, while lengthy or sufficient time increases the odds of having better results. Techniques to counteract these influences can be employed, for example, taking an inventory of all the influences that may be in play, having a healthy skepticism about the information you are assessing, or discussing observations with others who may not be experiencing the same influences at that time before making a decision. Cognitive capacity also plays a role in situational awareness. Each individual has boundaries regarding how much data he or she can process and make sense of at any given time. The limitation creates some variability in the level of awareness and follow-on decision-making. Bad input results in bad output. In process work, situational awareness requires the ability to accurately identify and anticipate the needs of the organization, team members, and other stakeholders. It involves taking risks when predicting future needs to design new or improved processes. While mapping processes for confirmed needs is fairly straightforward, having accuracy for predicted needs can be more challenging. A recommended technique for addressing predicted needs includes low-tech penciling in of optional process flows that attempt to test prediction validity with as many stakeholders as possible before committing to a course of action. This method helps to reduce human error and bias introduced in inaccurate situational awareness, which is typically introduced when processes are designed in isolation and reality can be distorted. A situation map that articulates the change management dimensions that need to be considered can also be created. The map can take the form of a dimensional chart such as pie, line, or bar charts; information cubes; histograms; radar charts; process/flowcharts; and decision trees. Frequent inspection and show-and-tell of maps during process improvement initiatives serve to bring forward alternate points of view (POV) that can be leveraged to build accurate and comprehensive situational awareness. The value of POV insight translates into right-sized and valid processes that fit the evolving business landscape. Continuous assessment and communication are powerful controls to ensure high-quality results.

Competency examples include

- Seeing the interrelationships between parts of the organization
- Knowing the reasoning behind key policies, practices, and procedures and seeking exceptions when needed to achieve goals

- Understanding internal and external politics and their impacts on the organization

- Aligning resources and maneuvering politics to solve problems or reach goals

- Understanding the organizational structure and reporting relationships

- Understanding guidelines that enable the business to operate effectively

- Recognizing the influence and impact of the organization's key stakeholders

- Maintaining an up-to-date understanding of current organizational topics and issues

- Creating and maintaining successful business partnerships by developing an understanding of business and social norms in other cultures

Process Improvement Tools and Techniques

The following section provides an introduction to a variety of *tools and techniques* that can be used during Process Improvement activities. No specific tool is mandatory, and any one may be helpful at various points throughout the Process Improvement lifecycle.

Process Mapping

Process Mapping is the act of visually describing the flow of activities in a process and outlining the sequence and interactions that make up an individual process, from beginning to end. These maps show how work is done (as-is state) or could be done (to-be state) between different functional areas within an organization. Process Maps not only detail the sequence of activities within a process, they also identify handoffs and help define roles and responsibilities. The purpose of process mapping is to better understand the activities performed by an organization. It involves the gathering and organizing of facts about the work and displaying them so that they can be questioned and improved by operators, managers, and other stakeholders. In many cases, process maps serve as a communication tool, a business-planning tool, and a tool to help manage an organization. Key elements of a process map include the inputs and outputs of a particular process (how inputs and outputs are handled and distributed), the steps or activities involved in executing the process (what is happening), any major decision points, and the departments involved in the processes execution (who is operating it).

Benefits of Process Mapping

Organizations and Process Improvement Teams use Process Mapping to

- Articulate how work gets done in an organization
- Show how work should be done in an organization
- Improve an organization's understanding of business and operational performance
- Give employees an opportunity to gain familiarity in a shared view
- Reveal issues and gaps that cause business and operational problems; help increase the probability of success and encourage a higher level of involvement and agreement
- Allow a person to visually illustrate and convey the essential details of a process in a way that written procedures cannot do; can replace many pages of words
- Influence and accelerate the "Opportunity Assessment" and "Design" phases of a project
- Aid in solving problems and making decisions
- Show processes broken down into steps, using symbols that are easy to follow and understand
- Show intricate connections and sequences easily, allowing for immediate location of any element of a process
- Help organizations understand the important characteristics of a process, allowing it to generate useful analytical data in order to derive findings, draw conclusions, and formulate recommendations
- Allow organizations to systematically ask many important probing questions that lead to development of a view on business process improvement

How to Create a Process Map

The following list outlines the steps to take when creating a Process Map:

1. Use boxes to illustrate the activities or steps that make up the process in question.
2. Draw a line with an arrowhead to show an input or an output associated with each step.
3. Label the inputs and outputs of each step where applicable so that stakeholders are able to see the transformation or value of each step.
4. Keep the sequence of activities moving from left to right and avoid backflow or backward arrows.

5. Avoid confusing intersections of flow lines by using intersection symbols.

6. Use a diamond-shaped box to indicate when a decision is made in the process.

7. Label the decision and the decision outcomes from any decision points in the process.

8. Draw horizontal swim lanes to represent the different departments that participate in the process.

9. Outline the customer of the process in the top swim lane.

10. If several areas jointly perform the same step, draw a hanging box so that it overlaps all of the areas involved.

11. Label each step in verb–noun fashion.

12. Add any necessary annotations and process components to the map so that it is an all-encompassing view of the process (i.e., systems, business rules, data elements etc.).

13. Use standardized shapes across all process flows.

Figure 8-1 provides a basic overview, with descriptions and meanings, of the most common process mapping shapes.

Common pitfalls when drawing process maps include

- Not involving the affected persons in their creation
- Not verifying the content outlined in a map with stakeholders once it is completed
- Omitting information
- Not having a method to update and maintain the process map
- Not having process maps readily available

Force Field Analysis

Force Field Analysis is a technique used to identify forces that may help or hinder achieving a change or improvement within an organization. It is a method for listing, discussing, and evaluating the various forces for and against a proposed change so that the forces that help can be reinforced and the forces that hinder can be reduced or eliminated. Having identified these, an organization can then develop strategies to reduce the impact of the opposing forces and strengthen the supporting forces. An effective team-building tool, it is especially useful when it is necessary to overcome resistance to change. When a change is planned, Force Field Analysis helps organizations see the effects on the larger business landscape by analyzing all of the forces impacting the change and weighing its pros and cons. Once these forces have been identified and analyzed, it is possible for an organization to determine if

Shape	Shape Name	Description of Shape
	Activity	Activity, action, or step. This is the most common shape in process maps.
	Sub-Process	A sub-process shape is a marker for a sub-routine or series of process steps that are formally defined elsewhere.
	Delay	The Delay flowchart symbol depicts any waiting period that is part of a process.
	Standard Line	Standard line connectors show the direction that the process flows.
	Dashed Line	This line is typically used when one step leads in to multiple activities and symbolizes an alternate path.
	Manual Input	Manual Input shapes are process steps where the operator is prompted for information that must be manually entered into a system.
	Start	The start shape shows the start point in a process and depicts a trigger action that sets the process into motion.
	End	The End shape shows the end point in a process and depicts a trigger action that ends the process.
	Decision	Indicates a question or branch in the process flow. Typically, a Decision flowchart shape is used when there are 2 options (Yes/No, Go/No-Go, and so on).
	Off-Page Connector	Off-Page Connector shows continuation of a process onto another page or into another process map. Connectors are usually labeled with capital letters or numbers to show matching jump points.
	On-Page Connector	On-Page Connectors are used to show a jump from one point in the process to another. Connectors are usually labeled with capital letters (A, B, AA) to show matching jump points.
	Document	The Document shape is for a process step that produces a document.
	Multi-Document	The Multi-Document is for a process stape that prodices multiple documents.
	System	A system symbol represents a step or action that involves a specific application, system or database. This shape can also represent any other technical system where data may be sent or stored.
	Annotation	Used to supply additional relevant information to the audience or user of a process map.
	Or	The Or Shape shows when a process diverges - usually for more than 2 branches. When using this symbol, it is important to label the out-going flow lines to indicate the criteria to follow each branch. Often, this shape is used in conjunction with the alternate path line.
	Display	Indicates a process step where information is displayed to a process operator or user.
	Key Performance Measure (KPM)	Indicates a Key performance indicator or point of measure in a process.
	Junction Point Arrow	Indicates the direction of the Process flow when multiple process activities lead to a single process activity

FIGURE 8-1 Process Mapping Shapes

a proposed change or improvement is viable. Several factors can be analyzed, including people, resources, attitudes, traditions, regulations, values, needs, and desires. Force Field Analysis helps identify those factors that must be addressed and monitored if change is to be successful. Forces that help an organization achieve change are called driving forces and any force that might work against the change is called a restraining forces. Other types of forces that can be considered in force field analysis include

- Organizational structures
- Vested interests
- Relationships
- Social trends
- Personal needs
- Present or past practices

Benefits of Force Field Analysis

Organizations and Process Improvement Teams use Force Field Analysis to

- Ensure proposed changes are seen from a variety of perspectives
- Enable people to think together about all aspects of making a change
- Enable an organization to see objectively whether a proposed product will have a receptive market
- Determine potential strategic alliances
- Build consensus within an organization

How to Conduct Force Field Analysis

The following list outlines the steps to take when conducting a Force Field Analysis:

1. Using adjectives and phrases, describe the current situation or problem as it is now and the desired situation as the vision for the future.
2. List all the driving and restraining forces for the change.
3. Discuss the key restraining forces and determine their severity.
4. Discuss the key driving forces and determine their strength.
5. Allocate a score to each force using a numerical scale, with 1 being very weak and 10 being very strong.
6. Chart the forces by listing, in strength scale, the driving forces on the left and the restraining forces on the right.
7. Explore the restraining forces and the best way to address them.

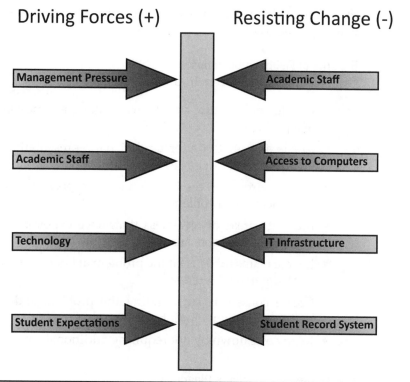

FIGURE 8-2 Force Field Analysis Template

8. Prioritize the driving forces that can be strengthened or identify the restraining forces that would assist with achieving desired state if they were removed.

9. Define the changes required in order to resolve the problem.

10. Identify priorities and produce an action plan.

11. Develop a Comprehensive Change Strategy.

12. Implement the required changes.

Figure 8-2 illustrates the basic format of a Force Field Analysis template.

Fishbone Diagrams

A *Fishbone Diagram*, also known as a Cause-and-Effect Diagram or Ishikawa Diagram, is used to visually display the potential causes of a specific problem or event. A Fishbone Diagram provides structure for problem-solving discussions and is used to help individuals find and cure root causes of a problem rather than its symptoms. It is particularly useful in group settings and for situations in which little quantitative data are available for analysis. Fishbone Diagrams are also used to assist in illustrating the relationships between several potential or actual causes of a performance problem.

Benefits of Fishbone Diagrams

Organizations and Process Improvement Teams use Fishbone Diagrams to

- Enable analysis that avoids overlooking any possible root causes of a problem
- Create an easy-to-understand visual representation of a problem, its causes, and categories of causes
- Focus a group on the big picture as to possible causes or factors influencing the problem
- Illustrate areas of weakness that, once exposed, can be rectified before causing more sustained difficulties
- Enable teams to focus on the problem at hand and not its history or related personal interests
- Create consensus surrounding the problem and its causes and build support for solutions
- Keep costs down by not requiring additional software

How to Create a Fishbone Diagrams

Following are the steps to take when conducting a Fishbone Diagram:

1. Agree on a problem statement.
2. Create the main fishbone and place the problem statement on the right-hand side.
3. Brainstorm the major categories of causes of the problem.
4. Write the categories of causes as branches from the main fishbone.
5. Continue to generate deeper levels of causes and add them as layers to each branch.
6. Analyze the diagram and its various symptoms.
7. Determine Root Causes.
8. Identify and prioritize solutions.
9. Create an Action Plan for resolution.

Figure 8-3 illustrates the basic format of a Fishbone Diagram.
The following list outlines some tips for crafting fishbone diagrams:

- Make sure that there is group consensus about the problem statement before beginning the process of building the fishbone diagram.

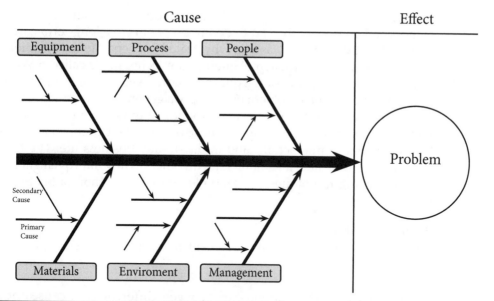

Cause   Effect

FIGURE 8-3   Format of a Fishbone Diagram

- Consider as much information about the problem, including what, where, when, and how often, in order to give the team as much background knowledge as possible.

- If appropriate, separate branches that do not contain a lot of information onto other branches.

- If appropriate, split branches that have too much information into two or more branches.

- Only use as many words as necessary to describe the various causes and effects.

- If ideas are slow to come out of a group, use the major cause categories as catalysts or triggers for discussion.

NOTE *The 5 Whys, Brainstorming, and Check Sheet techniques can be used to generate the causes and subcauses during a Fishbone exercise.*

---

5 Whys

5 Whys is a question-asking technique used to explore the root cause of a particular defect or problem. It involves looking at a problem and asking several times why it occurred. The first answer prompts another why, the answer to the second why will prompt another, and so on. This is critical as symptoms often mask the causes of problems. As with effective incident classification, basing actions on symptoms is the worst possible practice.

Effective use of the technique will define the root cause of any nonconformances and subsequently lead to definition of effective long-term corrective actions. By repeatedly asking why, teams can separate the symptoms from the causes of a problem. Although the technique is called 5 Whys, teams will often have to ask the question more than five times in order to find the true cause of a problem (5 is simply a general rule of thumb). Conversely, it may take teams only a few rounds of questioning to get to the root cause of an issue. 5 Whys is most useful when problems involve human factors or interactions and in day-to-day business life. The technique can be used alone of combined with other quality improvement and troubleshooting techniques, for example, within or without a Six Sigma project.

Benefits of 5 Whys

Organizations and Process Improvement Teams use the 5 Whys to

- Quickly identify the root cause of a problem
- Determine the relationship between different root causes of a problem
- Avoid the use of statistical analysis or advanced mathematics
- Foster and produce teamwork and teaming within and without the organization
- Avoid high costs and reduce the amount of setup

How to Complete the 5 Whys

The following list outlines the steps to take when conducting a 5 Whys evaluation:

1. Assemble a team of people knowledgeable about the area of nonconformance.
2. Write a description of the problem; describe it as completely as possible and come to an agreement with the team on its definition.
3. Ask the group why the problem is occurring and write the answer down underneath the problem description.
4. If the answer doesn't identify the root cause of the problem, then ask why again and write that answer down.
5. Continue to do this until the team is in agreement that the problem's root cause has been identified.

Figure 8-4 illustrates the basic format of a 5 Whys audit sheet.

NOTE *The 5 Whys technique can be used individually or as part of the fishbone diagram exercise.*

FIGURE 8-4 Format of a 5 Whys Audit Sheet

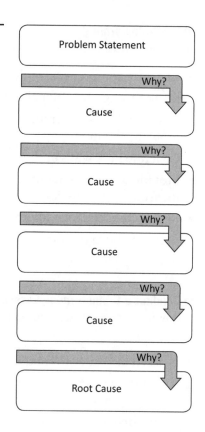

Brainstorming

Brainstorming is a technique used to generate a large number of creative ideas within a group or team of people. It can also be used to identify solution alternatives and to obtain a complete list of items for a particular endeavor. The Brainstorming technique is intended to create an atmosphere in which people feel uninhibited and free to propose solutions to problems without criticism, evaluation, or judgment. Each group member, in turn, can put forward ideas that might normally seem unconventional or outlandish concerning the problem being considered, thus ensuring all possible options are considered for subsequent analysis. Ultimately, the process continues until no more ideas are forthcoming and the chances for originality and innovation are increased. Brainstorming in a Process Improvement environment is primarily used to

- Identify problem areas
- Identify areas for improvement
- Design solutions to problems
- Develop action plans

Benefits of Brainstorming

Organizations and Process Improvement Teams use Brainstorming to

- Generate comprehensive lists of ideas or thoughts
- Break through traditional thinking about a problem
- Ensure all team members are involved and enthusiastic about problem solving
- Generate new ways of thinking
- Provide an environment for building on new ideas while staying focused on the mission, task, or problem at hand
- Reduce the tendency to prematurely discard new ideas
- Facilitate team building
- Encourage team-oriented problem solving

How to Conduct a Brainstorming Session

The following list outlines the steps to take when conducting a Brainstorming session:

1. Select a facilitator to conduct the Brainstorming session.
2. Write down the central question, topic, or purpose of the brainstorming session and ensure participants agree on intent.
3. Have each member of the team or group, by turn, suggest ideas one idea at a time.
4. Note each suggestion.
5. After one go-round of all members, repeat another cycle of idea gathering.
6. Anyone who has run out of ideas may pass their turn until all members have passed on their turn, indicating that idea generation has been exhausted.
7. Review ideas for clarity and remove any duplicates
8. Clarify ideas that were not evaluated if not clear.

Common pitfalls when conducting Brainstorming efforts include

- Not establishing a clear objective for the brainstorming session to ensure the group is focused
- Allowing members to shout ideas randomly without a structured round-robin approach
- Not ensuring participants offer only one idea at a time
- Allowing one individual to monopolize the session

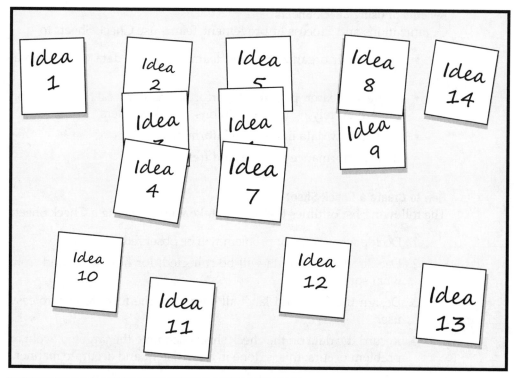

FIGURE 8-5   Example Brainstorming Session Output

Figure 8-5 illustrates a sample Brainstorming session output.

NOTE *Brainstorming can be used in conjunction with the fishbone diagram technique.*

Check Sheet

A *Check Sheet* is used to collect and structure data so that decisions can be made based on facts, rather than anecdotal evidence. Data are collected and ordered in a table or form by adding tally or check marks against problem categories. It allows a group to record data from either historical sources or as activities are happening so that performance patterns or trends can be detected. A Check Sheet presents information in an efficient, graphical format that registers how often different problems occur and the frequency of incidents that are believed to cause problems. Check Sheets are primarily used to

- Identify what is being observed in a particular process or environment
- Collect data in a easy and simple fashion
- Group data in a way that makes it valuable and reliable to stakeholders

Benefits of Using Check Sheets

Organizations and Process Improvement Teams use Check Sheets to

- Enable their organization and teams to collect data with minimal effort
- Create a clearer picture of performance with each observation rather than relying on the opinions of stakeholders
- Convert raw data into useful information
- Make performance patterns and trends

How to Create a Check Sheet

The following list outlines the steps to take when creating a Check Sheet:

1. Decide what event or problem will be observed.
2. Decide when the data will be collected, for how long, and from what sources.
3. Design the form and label all spaces on the form to ensure easy use.
4. Record the data on the check sheet each time the targeted event or problem occurs; this is done in a consistent and accurate manner.

Figure 8-6 illustrates a sample Check Sheet.

Pareto Analysis

Pareto Analysis, also known as a Pareto Diagram or Pareto Chart, is a bar chart that is typically used to prioritize competing or conflicting problems

Paint Job Quality Control Checklist

Job : 61945
Inspector: Adam Fraiser

Problem	Frequency
Chip	ⳋⳋ ⳋⳋ |||
Bubble	|||
Run	ⳋⳋ	
Scrape or Scratch	
Inadequate Coverage	ⳋⳋ ⳋⳋ ⳋⳋ ||
Other	

FIGURE 8-6   Sample Check Sheet

or issues so that the resources allocated to the problem offer the greatest potential for improvement by showing their relative frequency or size in a bar graph. A Pareto diagram or chart pictorially represents data in the form of a ranked bar chart that shows the frequency of occurrence of items in descending order. These charts are based on the Pareto Principle, which states that 80 percent of the problems come from 20 percent of the causes. The Pareto analysis technique is used primarily to identify and evaluate nonconformities, although it can summarize all types of data and is the diagram most often used in Process Improvement Dashboards. Pareto Analysis can also be used in any general situation where you want to prioritize items.

Benefits of Pareto Analysis

Organizations and Process Improvement Teams use Pareto Analysis to

- Breakdown a big problem into smaller pieces
- Identify most significant factors
- Show where to focus efforts
- Allow better use of limited resources
- Set the priorities for many practical applications
- Allow better use of limited resources

How to Conduct Pareto Analysis

The following lists outlines the steps to take when conducting a Pareto Analysis:

1. List all items that require analysis and charting.
2. Determine how many items must be measured to build a representative chart.
3. Measure the elements, using the same unit of measurement for each element.
4. Calculate the percentage for each element out of the total measurement.
5. Plot the bars, with the highest bar on the left.
6. Analyze the results.
7. Work on solving the most important element first.

Figure 8-7 illustrates the basic format of a Pareto Chart.

NOTE *Pareto Analysis is frequently used to analyze the ideas, issues, and topics from a brainstorming session.*

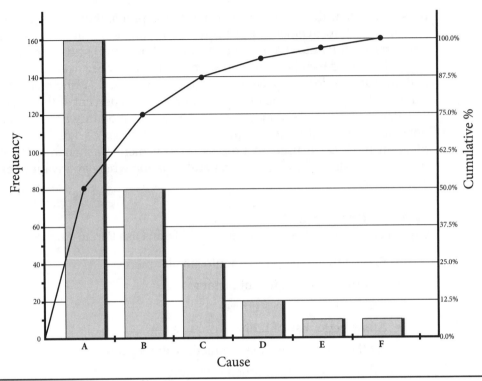

FIGURE 8-7 A Basic Pareto Chart

5S

5S is a structured technique used to methodically achieve organization, standardization, and cleanliness in the workplace. It is used to reduce waste and optimize productivity by maintaining an orderly workplace and uses visual cues to achieve consistent operational results. Implementation of this method helps organizations organize themselves and is typically one of the first Process Improvement techniques that organizations implement. A well-organized workplace can result in more efficient, safer, and more productive operations and can boost the morale of an organization's workforce. The 5S pillars or practices are based on five Japanese words: Sort (Seiri), Set in Order (Seiton), Shine (Seiso), Standardize (Seiketsu), and Sustain (Shitsuke) (Figure 8-8). This technique encourages workers to improve their working conditions and then reduce waste, eliminate unplanned downtime, and improve processes.

Benefits of using 5S

Organizations and Process Improvement Teams use 5S to

- Improve safety
- Raise employee morale

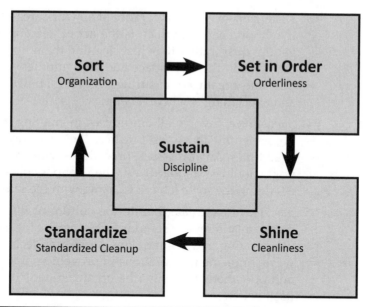

FIGURE 8-8   The Five 5S Practices

- Identify problems more quickly
- Develop control through visibility
- Increase product and process quality
- Promote stronger communication among staff
- Empower employees to sustain their work area
- Reduce set-up times
- Reduce cycle times
- Increase floor space
- Lower safety accident rate
- Reduce wasted labor

How to Use 5S

Following are the steps to take when using 5S:

1. *Sort:* Eliminate all unnecessary tools, parts, and instructions; keep only essential items and eliminate what is not required from all work areas.

2. *Set in Order:* Arrange all tools, parts, supplies, equipment, manuals, and instructions in a way that the most frequently used items are the easiest and quickest to locate. This is done to eliminate time wasted in obtaining the necessary items for use in a process or operation.

3. *Shine:* Improve the appearance of all work areas. Conduct preventive housekeeping, which is the act of keeping work areas from getting dirty, rather than just cleaning them up after they become dirty. Clean the workspace and all equipment and keep it clean, tidy, and organized, ensuring everyone knows what goes where and everything is where it belongs.

4. *Standardize:* Involve all employees in creating best practices and implementing those best practices the same way, everywhere, and at all times. Any processes, procedures, work areas, and other settings should be standardized wherever possible so that all employees doing the same job are working with the same tools.

5. *Sustain:* Indoctrinate 5S into the culture of the organization and maintain and review standards. Maintain a focus on this new way of operating and do not allow a gradual decline back to the old ways. Implement continuous improvement into general practices and operations to ensure the organization is continually making itself better.

There are three other items that organizations often include into 5S practices: safety, security, and satisfaction. Although these are not traditional 5S practices, they have evolved into what many organizations believe are critical to sustaining success. These items are

- *Safety:* Although safety is inherent in all 5S practices, some organizations believe that by explicitly stating safety as a practice it better promotes the practice and constantly reminds employees of its importance.

- *Security:* Security is another practice often added by organizations in order to identify and address risks to key business categories including materials, intellectual property, human capital, brand equity, and information technology.

- *Satisfaction:* Satisfaction is a pillar that can also be included. Employee satisfaction and engagement in Process Improvement activities ensure that any improvements made will be sustained and improved upon over time.

Affinity Diagram

An *Affinity Diagram*, also called an Affinity Chart, is a graphical tool used to organize ideas generated in brainstorming or problem-solving meetings. With this method, ideas are grouped into meaningful categories called affinity sets in order to understand the essence of a problem or performance issue. These categories tie different concepts together with an underlying theme and help clarify issues for team members so that better solutions can be designed.

Benefits of using Affinity Diagrams

Organizations and Process Improvement Teams use Affinity Diagrams to

- Help teams sift through large amounts of data
- Encourage creativity from team members involved in the improvement effort
- Break down communication barriers
- Encourage ownership of actions that result from the brainstorming session
- Overcome analysis paralysis, which is often brought on by having too many options or ideas for improvement

How to Create an Affinity Diagram

The following list outlines the steps to take when creating an Affinity Diagram:

1. Describe the issue under consideration or investigation.
2. Brainstorm ways of rectifying the issue.
3. Record each idea.
4. Sort the ides into related groupings or categories.
5. Create group consensus.
6. Create a description or summary for each category.

Figure 8-9 illustrates the basic format of an Affinity Diagram.

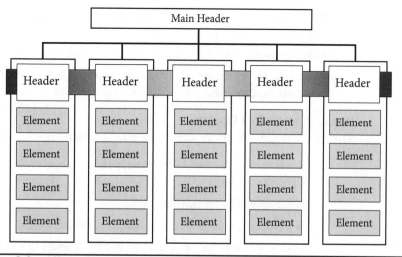

FIGURE 8-9   Affinity Diagram

Common pitfalls when gathering and grouping ideas include

- Not using neutral statements to describe problems, categories, or ideas
- Generating too many ideas; a typical threshold is fewer than 50
- Allowing participants to continually move ideas between categories
- Not allowing some ideas to stand on their own
- Not spending enough time describing the categories of ideas

Tree diagram

A *Tree Diagram*, also known as a Systematic Diagram, Analytical Tree, or Hierarchy Diagram, is used to break down broad categories of content into finer levels of detail. The diagram starts with a single goal that is broken down into detailed actions that must be completed in order to achieve the goal. Tree Diagrams are excellent tools to use

- When developing actions to carry out a solution or other plan
- When analyzing processes in detail
- When probing for the root cause of a problem
- When evaluating implementation issues for several potential solutions
- After an affinity diagram or relations diagram has uncovered key issues
- As a communication tool to explain details to others

Benefits of Using Tree Diagrams

Organizations and Process Improvement Teams use Tree Diagrams to

- Encourage team members to expand their thinking when developing solutions
- Allow all team members to check the link and completeness of every level of a plan
- Help move teams from theoretical to real-world solutions
- Reveal the complexity involved with the achievement of a goal or solution, ensuring projects and changes are truly manageable

How to Create a Tree Diagram

The following list outlines the steps to take when creating a Tree Diagram:

1. Develop a statement of the goal, project, plan, problem, or whatever is being studied.
2. Assemble an appropriate team.
3. Develop the major tree headings (subgoals).
4. Ask questions that will lead to the next level of detail.
5. Brainstorm all possible answers.

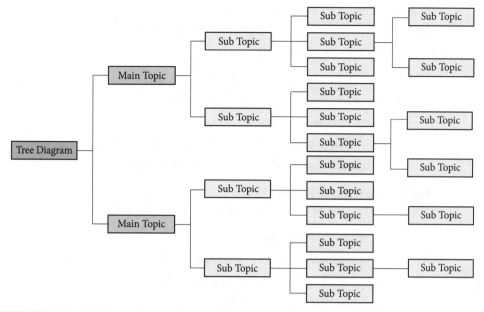

FIGURE 8-10 Tree Diagram

6. Do a sufficient check of the entire diagram to ensure flow and completeness.

7. Consider proposed changes and modify the tree as needed.

8. Develop a change strategy.

9. Implement changes.

Figure 8-10 illustrates the basic format of a Tree Diagram.

A3 Report

The *A3 report* was developed as a decision-making tool in the 1980s by Toyota Motor Corporation. It refers to a European paper size that is used to encompass various pieces of important information on a single page. The A3 document provides a structure and a consistent format for communications and problem-solving methods. Many companies use the A3 report when planning improvements to processes. Traditionally A3 reports were used to document and show the results from the PDCA (Plan-Do-Check-Act) cycle on a single page. The A3 format is now commonly used as the template for three types of reports:

- Proposals
- Status
- Problem-solving

Benefits of using A3 Reports

Organizations and Process Improvement Teams use A3 Reports to

- Provide a logical thinking process
- Clearly present known information objectively
- Focus on and share critical information
- Align efforts with strategy/objectives
- Provide a consistent approach throughout the organization
- Provide a powerful problem-solving process

How to Create an A3 Report

Following are the steps to take when creating an A3 Report:

1. Identify the problem or need.
2. Understand the current situation/state.
3. Develop the goal statement or target state.
4. Perform root cause analysis.
5. Brainstorm/determine countermeasures.
6. Create a countermeasures implementation plan.
7. Check results and confirm the effect.
8. Update standard work.

Figure 8-11 illustrates the basic format of an A3 Report.

Scatter Diagram

A *Scatter Diagram*, also called a scatter plot or X–Y graph, is used to analyze relationships between two variables. One variable is plotted on the horizontal (X) axis and the other is plotted on the vertical (Y) axis. The pattern of their intersecting points can graphically show relationship patterns. Most often a scatter diagram is used to prove or disprove cause-and-effect relationships.

If the variables are correlated, the points will fall along a line or curve. The better the correlation, the tighter the points will hug the line. Scatter Diagrams are most commonly used

- When you have paired numerical data
- When trying to determine whether the two variables are related
- When trying to identify potential root causes of problems
- After brainstorming causes and effects using a fishbone diagram in order to determine objectively whether a cause and an effect are related

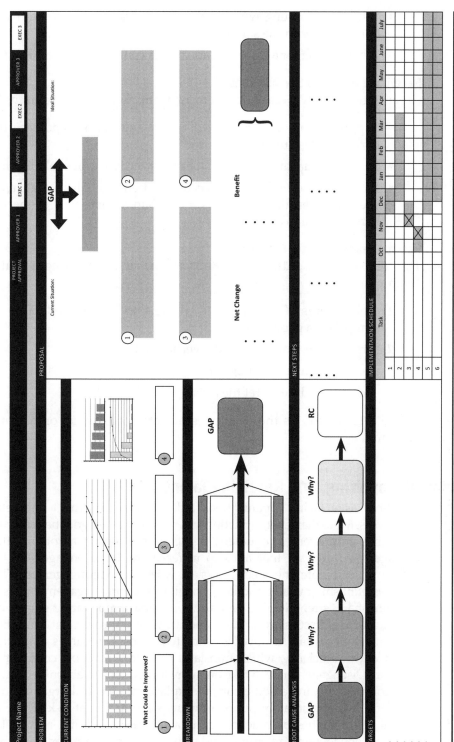

FIGURE 8-11 A3 Report

247

- When determining whether two effects that appear to be related both occur with the same cause
- When testing for autocorrelation before constructing a control chart

Benefits of Using Scatter Diagrams

Organizations and Process Improvement Teams use Scatter Diagrams to

- Supply data in order to confirm a team's theory or hypothesis that two variables are related
- Provide visual and statistical means to verify various relationships
- Provide a good follow-up to cause-and-effect analysis that can be used to determine if there is more than a consensus connection between causes and effects

How to Create a Scatter Diagram

The following list outlines the steps to take when creating a Scatter Diagram:

1. Collect pairs of data where a relationship is suspected; 50 to 100 samples is a good range.
2. Draw a horizontal axis and a vertical axis.
3. Plot the data on the diagram.
4. Analyze and interpret the data.

Figure 8-12 illustrates the five types of data correlations on a scatter diagram.

Building, Recognizing, and Retaining Talent

A critical issue facing organizations today is how to attract and retain employees. More than ever, attracting and retaining talent goes beyond compensating employees adequately. It includes building a robust talent management program that aims to increase employee engagement and reduce or eliminate the steep costs of workforce disengagement. This includes publishing values and key behaviors, securing engagement, fostering mindshare, strengthening organizational goals, motivating employees, and facilitating discussion of ideas and areas of improvement. In today's economy, unnecessary turnover is estimated to cost companies as much as 30 percent in efficiency, which translates to a very real cost containment problem. Disengagement can take on many forms, such as absenteeism, low attention to detail, increased errors, workplace conflict, and organizational bloat and it can ultimately put a company's reputation at risk. There are several key activities or practices that can be used to build and sustain employee engagement, retention, and satisfaction and to prevent employee

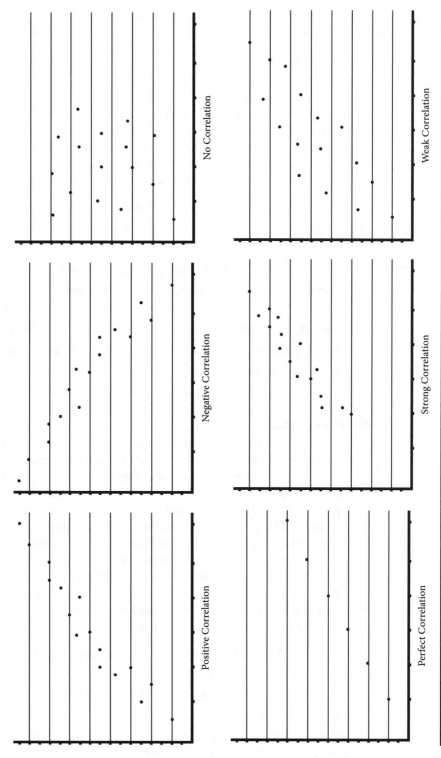

FIGURE 8-12 Scatter Diagram

249

disengagement and attrition. This section outlines several key practices that organizations can use to build an effective talent management program.

Valuing Workforce Members

An organization's success depends increasingly on an engaged workforce that has a safe, trusting, and cooperative work environment. Successful organizations capitalize on the diverse backgrounds, knowledge, skills, creativity, and motivation of their workforce and partners. Valuing the workforce means committing to their engagement, satisfaction, development, and well-being. This involves more flexible, high-performance work practices that are tailored to varying workplace and home life needs. Organizations can do this by

- Demonstrating leadership's commitment to employee success
- Providing recognition that goes beyond the regular compensation system
- Offering development and progression within the organization
- Sharing organizational knowledge so that employees can better serve customers and contribute to achieving strategic objectives
- Creating an environment that encourages risk taking and innovation
- Creating a supportive and diverse environment for the workforce
- Offering work from home policies
- Scheduling regular social events for employees and staff members

Continuous Learning

Employee training is an essential activity for organizations. It shows that the organization invests in its employees and is also a key component in promoting loyalty and commitment. The primary objective of training is to provide all personnel, suppliers and customers, with the skills needed to effectively perform process activities and to build this concept directly into the improvement framework. This practice enables continuous learning within the organization and promotes improvement and process-oriented thinking. Formal individual and team training plans are an excellent method for keeping track of learning goals and development paths of organizations, teams, and individuals. Training can be proactive or just-in-time in its approach, depending on the needs and financial means of the department or organization.

Other methods of continual learning are

- Learning from past mistakes and successes (conducting retrospectives and documenting lessons learned)
- Creating communities of practice for learning
- Hosting internal learning sessions
- Formally training employees on internal values and practices as well as industry skills and methods

Continuous Learning

- Encourages knowledge creation
- Fosters knowledge sharing
- Builds innovation and best practices
- Helps to discover hidden knowledge and expertise
- Reduces relearning
- Promotes a sense of team
- Contributes to business success
- Improves performance
- Increases motivation

Rewards and Recognition

Rewards and Recognition programs honor both individuals and teams who go the extra mile to service their departments and customers. Having a rewards and recognition program in place lets valued employees know that their contributions are important and that their efforts are appreciated. By creating a culture of recognition, employees become more engaged, which leads to higher productivity, motivation, and engagement in improvement activities. Examples of Awards Programs that Process Improvement Organizations often adopt are

- *Spot Awards:* Spot Award Programs recognize individuals for their special effort within their department or an improvement initiative on an impromptu basis. This award can be presented on the spot when a manager notices an employee doing something worthy or, for a larger impact, at a staff meeting so the employee can be recognized by peers.
- *Team Achievement Awards:* The Team Achievement Award recognizes teams within or across departments that meet the following criteria:
  - Complete an important process improvement step or milestone
  - Provide significant positive impact on a new or existing activity, process, or project
  - Promote and demonstrate team collaborative behaviors and embrace the Process Improvement framework
  - Demonstrate sustained effort

Although these are widely used methods of recognition, not all employees respond to rewards in the same manner. Organizations must always ensure that the recognition provided is inline with an employee's comfort level.

Performance Management

Performance Management is the process by which companies ensure alignment between their employees and company and department goals and priorities. Performance Management is a critical element in proper Talent Management and is designed to promote interaction and feedback between management and employees, establish expectations for individual work performance and to serve as a foundation for rewarding top employees. Performance Management systems require a structured method of communication between all levels of an organization. Companies that require and promote this feedback loop can learn and grow from the information acquired. The benefits of performance management include reliable measurement of productivity, benchmark for performance, clear understanding of the employees objectives and goals, ability to provide focused feedback, and providing management with an ability to pinpoint performance problems. In addition, performance management ensures targets and goals are transparent and well documented.

The performance management cycle is extremely important as this is a method used to evaluate desired behavior against actual performance. The cycle includes the following seven steps:

1. Define goals and objectives.
2. Ensure performance expectations and processes are clear.
3. Monitor and evaluate performance.
4. Provide feedback and adjust processes or performance as needed.
5. Conduct formal performance appraisal sessions; these can be held daily, monthly, or yearly.
6. Regularly request feedback from peers, leaders, and customers.
7. Conduct leadership calibration and employee potential sessions.

An organization that does not properly implement and support a performance management system may not experience all of the possible benefits of increased communication and workforce development.

Other Talent Management Practices

Following is a list of other talent management practices:

- *Compensation:* Compensation remains a key factor in attracting and retaining top talent. To attract and retain employees, compensation needs to be competitive and fair. In this economic time period, smart employees are doing their homework on this topic, and employers need to be prepared to pay a reasonable wage for top-tier talent.
- *Benefits:* Employees and employers often rank benefits as a top reason why employees stay with an organization. Offering a broad range of

benefits to an increasingly diverse workforce shows that the employer is responsive and aware of the needs of each employee. Allowing for a range of benefits for your employees is a great retention tool.

- *Healthy Workplace Initiatives:* Healthy Workplace Initiatives are a new trend in the workplace environment. Promoting a healthy workplace is becoming a key factor in keeping healthcare costs and workplace injuries down and in attracting talent. These initiatives can include promoting a clean and safe work environment to support healthful lifestyles, fitness goals, and healthy diets. Free fitness and lifestyle programs are key to attracting top-tier staff.

- *Work–Life Flexibility Programs:* Work–Life Flexibility programs recognize that employees have important family obligations that compete with and sometimes take precedence over their workplace commitments. Dependent care leave, childcare subsidies, eldercare programs, counseling and referral, flexible working hours, and transportation benefits allow people to strike a more meaningful and potentially less stressful balance between obligations at work and home.

Chapter Summary

In this chapter, we reviewed the Process Improvement skills, competencies, and techniques needed take a systematic approach to optimizing business processes toward an efficient and successful outcome. Clearly defined aptitudes help to drive better selection of people, provide a better understanding of the development required, and provide a more objective performance management system. Having aptitudes explained in behavioral terms provides the basis for a common understanding across a team of not only what is required to succeed but also how people who are successful conduct themselves. The investment in learning and building the aptitudes presented in this chapter are believed to provide a significant return in results and value to organizations. Even minor improvements can have a meaningful impact in achieving the right results such as improving customer experience, improving employee satisfaction, reducing waste, and increasing profitability. In this chapter we also learned that

- The Process Improvement Aptitudes Dictionary outlines the underlying characteristics that define the patterns of behavior required for Process Improvement professionals to deliver superior performance in their role.

- A competency is the ability to meet complex demands by drawing on and mobilizing psychosocial resources, such as skills and attitudes, in a particular setting.

- A skill can be thought of as an individual's ability to perform tasks or solve problems, while techniques are the various ways of carrying out a particular task.

- There are 10 skills and 10 competencies that are considered essential for ensuring the success of individuals involved in Process Improvement efforts. When combined with various techniques, these skills and competencies can collectively deliver process excellence.

- Change Management is widely considered the most critical skillset when conducting Process Improvement efforts.

- The 10 core Process Improvement skillsets are Facilitation, Change Management, Project Management, Time Management, Analytical Thinking, Negotiating, Decision Making, Communication, Training, and Coaching.

- The 10 core Process Improvement competencies are Building Partnerships, Collaboration, Credibility, Flexibility, Initiative, Business Knowledge, Sound Judgment, Resilience, Strategic Implementation, and Situational Awareness.

- The tools and techniques most commonly used in Process Improvement are:
  - Process Mapping
  - Force Field Analysis
  - Fishbone Diagrams
  - 5 Whys
  - Brainstorming
  - Check Sheets
  - Pareto Analysis
  - 5S
  - Affinity Diagrams
  - Tree Diagrams
  - A3 Reports
  - Scatter Diagrams

- There are many proactive Talent Management strategies an organization can put in place to promote employee retention, even during challenging economic periods. These include implementing formal performance management, valuing workforce members, training and educating employees and customers, and rewarding and recognizing achievement.

Chapter Preview

Chapter 8 established the aptitudes and tools needed to be successful in Process Improvement work. In Chapter 9 real business scenarios that draw upon all of the information presented in the Process Improvement Handbook are explored. Through case examples and personally identifiable situations in which many of us find ourselves, this chapter offers observations about areas that are worthy of improvement and discusses the outcomes commonly involved in these situations.

CHAPTER 9

Case Examples

This chapter provides case examples and guidance on how to avoid common pitfalls associated with Process Improvement implementation. It is built around real-life events that point out the value of positive, consistent, and reliable service experiences and demonstrates examples of how improvement can be made using a process improvement framework such as *The Process Improvement Handbook*. The case examples outlined also provide opportunities for readers to develop and enhance application skills. Each case involves the integration of content across all chapters of the handbook, and presents issues encountered in social work settings related to the practice and execution of the Process Improvement knowledge areas. These case examples were selected to illustrate specific issues and provide insights for practitioners as they execute process design and improvement efforts. The goal is to include sufficient information to highlight lessons while allowing practitioners to judge whether these lessons could apply to their context. This chapter is organized around the following case examples:

- *Recognize What Failure Is:* This is a study in evolving customer expectations and the cost of keeping a customer.
- *Achieve Customer Delight:* This case example explores consistency of experience as a driver for achieving customer delight.
- *What Can Go Wrong, How to Make IT Right:* In this example, reflections of implementation flexibility and agility are provided using the process-oriented landscape.
- *Remove Roadblocks:* A review of the value of process measurement is presented in this case example.
- *Reward and Celebrate:* This case example presents results through social engineering and reinforcement.
- *30 Days Gets Results:* The strategic advantage to process visibility is discussed in this example.

Recognize What Failure Is

The Challenge

If you are like most people, you unconsciously interact with business processes every day but get frustrated when you are abruptly confronted with a process that doesn't work. Frustration is a natural response when you experience a mismatch in expectations and the service provided. Process work often focuses on achieving an efficient flow of work from one step to another or has the intent to save money and time. These objectives are clearly important—but alone they are incomplete and can breed friction between expectations and delivery. Planning the human interaction experience in process design is paramount to achieving a smooth interaction experience. Below is an example of an interaction experienced in 2011 that illustrates how a mismatch in process and expectations impacts business revenue and profitability and should be taken seriously.

The Solution

I had been an Amazon.com customer for more than 10 years when I lived in Michigan before moving to Ontario, Canada, to build a Continuous Improvement organization. Bright-eyed and eager to give 20 copies of Geary A. Rummler and Alan P. Brache's text *Improving Performance: How to Manage the White Space in the Organization Chart* as gift to build team unity, I turned to Amazon's Canadian website prepared to spend approximately $800.00 on the bulk order. Part of the attraction of buying from Amazon was based on free shipping and consistently positive shopping experiences in the past. The book was not listed as available to order from the Canadian site. Determined to get fast access to the book, I decided it would be worth the expense to import the books to Canada from the United States site so that I could give the gifts to my team at the next big meeting, which was in a week. I successfully ordered the books and opted for expedited delivery to Canada, which added about $15.00 to the total.

About four days later a big Amazon shipping box was at my doorstep when I returned home from work. The box had no visible damage, and when I opened it I found lots of brown packing paper bunched up at opposite ends of the box and two stacks of 10 books. Perfect. Or so I thought. I pulled the first book out and started flipping through the pages, enjoying the new book and imagining how happy I was about to make 20 employees with the gift. I set the book down and began to pull each copy out of the box and stack them on my desk. Then I noticed something unexpected. There was a black mark on the bottom of each book that extended from the top of the hard cover to the back along the pages. I looked closer. Ink had rubbed off on each book, essentially smudging the top hard cover. I was not impressed and didn't think the books looked new at all. I began to imagine what I would feel like if someone who was my new leader handed me a book that looked like this and told me that it was a gift. I would think the person gave me a used book and would feel like it was not much of a gift at all.

Being a reasonable person who just invested more than $800.00, I hopped onto Amazon.com and found a fantastic set of customer support tools that let me upload a photo of the problem and explain the problem for their review. In my message, I asked Amazon to immediately ship out replacement books and provide return shipping for the defective books. After about 10 clicks, I noticed a note that said if you want to return your shipment, there was another process to follow. I clicked that link, filled in all the information *again*, and hit submit. I immediately received an RMA number (return to manufacturer authorization number) and instruction to take the box to my local mail service and pay for the return. The instruction noted that Amazon would credit my account after they received the goods. I was disappointed and frustrated. If I did what I was instructed, I wouldn't have new books, plus I had all the bother of driving to the local mail service and paying even more money to return Amazon's books. I felt betrayed by what I perceived was a broken promise. I decided that using RMA the way they wanted me to wasn't good enough. I filled out another form asking that they contact me and provided my home phone number. In less than a minute, my home phone rang. I answered, and amazingly it was an Amazon employee asking me about the issue.

They already had a lot of my information from the forms I had filled out, but they did not have the pictures that I had uploaded. The employee said, "I don't have access to that information." So, I described the problem in as much detail as I could and told him that what would make me happy was "new" books that didn't have smudges on them. Also, I wanted them to send out the books immediately and to have a courier pickup the box of defective books from my front step. The employee was very polite and professional and clicked away as I talked. Eventually he informed me that because I lived in Canada, he could not accommodate my request due to some problem with getting the goods back through customs. I was able to order from Canada, but I wasn't able to return from Canada? This made no sense.

Things escalated as my frustration got the better of me. I told the representative that I was just going to throw the books away if they wouldn't take them back and shop locally and never shop at Amazon again. Whatever their process was, it sure wasn't helping me and I wasn't going to be taken advantage of. Then something remarkable happened. He asked me not to destroy the books and asked that I donate them to a school or charity in my area that could use them. He said he would expedite a new shipment out immediately in recognition of my long-standing loyalty to Amazon. I felt my frustration dip, and rational thought returned. How will I know that the next shipment won't have the same problem I asked? He said he would ensure that the warehouse would be notified of the problem and that they would make sure the quality was up to their standards.

Three days later the replacement box arrived. Twenty beautifully stacked books, just like the first shipment. As I opened the box, I took a very close look for the defect I had found the last time. Exactly the same marks and smudges on every book. Frustration! I carried the entire box of books into work the next day and plunked them down on a table in from of my team. I retold the entire story of my intent and excitement to share the fantastic book that I believed would change their lives. And then I told them the very interesting lesson I learned from my expensive Amazon lesson.

The Result

It is worth noting that Amazon did a lot of things right in their business process. They were attentive to me as a customer, they had many interactions thought out, and self-service was available on their website. They acted with integrity when trying to resolve the problem—and I would say did better than any company I had ever had experience with. They did not, however, have a complete understanding of their global business customer's needs, there were some gaps in service processes for people like me, and they lost money in their attempt to correct the problem. Perhaps mine is a fringe case for Amazon that doesn't affect their overall operation and they aren't overly concerned that this happened to me. However, I haven't shopped with Amazon Canada or US because they didn't meet my expectations, which ironically they had raised based on 10 years of fantastic shopping experiences. In the past, I spent a few thousand dollars each year at Amazon for business and personal purchases. Now I spend that money at a local store where I can inspect the books I'm buying. Amazon lost me as a customer but gave me greater appreciation for the importance of completeness in business process design that will always work for the customer. It is exactly this appreciation that drove me to devote so much attention to the process ecosystem and process-oriented architecture. This case is written through experience from author Tim Purdie.

Case Summary

In this situation there was a conflict between the consumer's expectations for a simple and effective method of returning products based on consistently simple and high-quality ease of making purchases. The process for purchase was very well tuned for optimal performance and volume of transactions. The consumer had learned over a lengthy period of time to be able to trust that every purchase would have the same attributes of quality and delivery that built brand loyalty and favor. While there was a reliable and "as-designed" return method for consumers to use, the fact that I received two deliveries with the exact same issue broke my trust. It appeared that no process was in place to address repeating delivery problems differently, and there were no instructions on how to resolve high-value customer issues separate from those of occasional customers.

There is an opportunity to apply the principles found in the process-oriented landscape and add this situation to the use-case library and address the process gap in order to avoid further erosion of the customer base. Resolving the original issue regarding how the blemished books moved out of the warehouse and were delivered to a customer needs to be explored more deeply in order to determine root cause and corrective process action. In addition, a larger gap in process coverage that needs further evaluation was uncovered. The volume of customers who may have received blemished books may be a very tiny number among all the retailer's transactions. However, the notion that a high-dollar spending and long-term customer could be lost so easily needs to be explored more closely. An improvement in the retailer's overall process would incorporate process performance monitors or alerts for this tier of customer who report back-to-back quality problems with their shipment. These situations should be routed differently for a "high-touch" service response. The investment in white-glove programs that handle priority customers in this way has proven to aid in customer retention and improve brand reputation for other organizations.

Achieve Customer Delight

The Challenge

Having had my fair share of bad retail experiences with products that needed to be returned, I was delighted when one experience actually met with the marketing hype. Over the years I have purchased a winter coat for my children every year. As kids grow quickly, the investment is usually small because the coat is intended to last only one season. However, almost without fail, every brand of coat I have purchased failed to keep my daughter warm and dry during the harsh Canadian winter, and I had experienced the anti-customer delight experience many times while dealing with the retail outlet I purchased from. Each time it left me swearing off ever buying that brand the following winter. This past winter I purchased a more expensive brand of winter coat for my 10-year-old daughter from a local chain sporting good store rather than the usual department stores I had shopped at in years past. I did this because I had two expectations—first, that a specialty store would have higher quality coats and, second, that they would have professional service representatives who would not only know the best product for my daughter's needs but they also would intimately pick a product that would be more lasting than we had purchased before. Two hundred dollars later, we walked out of the store, my daughter happy with her purple coat and me happy that for once she would be warm and the coat wouldn't fall apart.

The Solution

Fast-forward two months. My daughter came home from school pointing out that her industrial-looking zipper was gaping open in the middle. How could this be? With a decidedly annoyed intent to read the store a piece of my mind, I rummaged through receipts to locate the one for the coat but came up short. As most of us have come to expect, no receipt means the store has no way of knowing that the coat was purchased from them and that it won't honor a return or replacement. While frustrating, the store does have the upper hand in this regard. However, my wife recalled that the store advertised their commitment to customer satisfaction on a large sign behind the cashier and called the store to ask what their policy was when a customer can't locate the receipt. To our delight, they told my wife they would do a straight exchange. Off to the store my daughter and I went with some hesitation that there would certainly be some hassle about the zipper to deal with. I explained our situation to the cashier, who quickly checked the store inventory and found the very last matching coat in my daughter's size and handed it over. No paperwork! Not even an entry into the computer. No, in fact, the cashier explained that this was very unusual for that brand of coat and that should anything go wrong with this one, I should bring it back. She further explained that if the zipper should fail on this coat, the manufacturer would ask me to take the coat to a local tailor and have them replace the zipper; I would pay the cost and bring the receipt to the retailer for reimbursement. Walking out of the store, I turned to my daughter and said, "This is why you spend a little more for quality—this never happened with your other coats!"

The Result

Reflecting on this example, there are a couple of points worth mentioning. First, because I had experience that taught me to be skeptical of retailers and their claims of customer satisfaction, I had a private expectation to be disappointed. I carried this with me into the sports specialty store. Second, I had the private expectation that paying more this time would remove the quality issue altogether and I wouldn't have a bad experience. I also had an expectation that I wouldn't have to keep my receipt handy as a result. These expectations were all in my head and had created a mismatch even before I walked into the store. There was an opportunity for the salesperson to inform me that included with the brand reputation was additional protection provided by the store because of their great exchange policy or their repair-at-their-cost policy. Unfortunately, they didn't take that opportunity; as a result, the anxiety I felt when I discovered the zipper had failed and the phone call to the store could have been avoided. That anxiety had tripped off the "I'll never shop there again" signal in my brain because I had been duped into spending more than I had in years past. That is a powerful consumer trigger that I am sure we have all felt. Alas, had the salesperson "sold" me on these qualities, I would never have had those residual emotions. Luckily for that store, their recovery was so quick and little time had passed from issue to resolution that I walked away with a happy perception of the entire experience and would very likely buy the next winter coat from that same store.

The lesson here is that customer delight is psychologically and emotionally driven by what is in the mind and experience of the customer. It is reasonable to expect retailers to look at the data on return rates of past purchases of specific brands and to measure follow-on sales by the same customer and realize something was wrong. They could have then interviewed, surveyed, and observed the problem and come up with processes that better inform customers. In addition, they could have handled returns more smoothly in an effort to make their customers happier, which would translate into future purchase loyalty, just like the expensive sporting goods retailer had clearly done. The reward to the business is many years of hundred-dollar purchases of coats.

Case Summary

A mismatch of expectations and delivery capability is the most common contributor to a lack of customer delight. Many organizations establish departments and dedicate staff to handle customer service concerns when the mismatch occurs. We argue that most mismatches could be avoided by implementing use-case scenarios for varying expectations that a customer would/could have and creating processes that guide the customer to a successful outcome. The transparency of the process is paramount as it sets an expectation that can match the planned delivery outcome.

In the winter coat return example, the value of a well-thought-out and advertised process that is followed can be observed. The retailer did what they said they would do. To the trained eye looking for process gaps, it was clear that thought had been put into the unhappy path of the process; meaning, when things don't work as expected, how is everything expected to work. This simple notion takes a concerted amount of time to plan for, model, and ensure the right consumer outcome while maintaining profitability. A lesson within this case example also emphasizes how meeting basic service expectations can be something a consumer would qualify as meeting their criteria for service delight. Grand gestures and discounting or financial incentives aren't always necessary—a well-designed process that works as advertised can do the trick.

What Can Go Wrong, How To Make IT Right

The Challenge

In September 2012 I took on a rip-and-replace project for a broken Customer Relationship Management (CRM) solution and a crippled business operation that needed life support. This example looks at a recent project in which a technology conversion from Microsoft CRM to Salesforce.com CRM took place. Beyond the very different business models of Microsoft and Salesforce.com, the company's underlying business processes needed to be reviewed from top to bottom in order to understand the gaps that would be solved by using the new tool. In addition, the new gaps that were created by the procedure changes needed to work with the new tool. Every technology project faces change resistance as things evolve. In this example, the project team chose to face that potential resistance by using use-case scenarios to flush out the business goals related to performing a specific process, as well as the expectations regarding how those steps would be sequenced. The use cases were essentially microprocesses.

The Solution

With a library of use cases, the team was able to organize their work around satisfying the use cases in the most lean and efficient way possible while taking into account the business rules, dependencies, and interconnection points with other systems and processes. In effect this formed an interwoven landscape of CRM processes. This level of process coverage for real-life business situations brought about a great deal of confidence not only in the project team but in the executive team that had taken on elevated risk in deciding to switch platform CRM solutions. The coverage translated to freedom to invent, adjust, and remove processes as needed in record time. During testing, when business users uncover new user scenarios, the project team used the newly built process landscape to assess the level of impact on the project and on the overall business in order to make decisions. Questions that were easily asked and answered included, Was the functionality proposed already solved by an existing process? Could an existing process be extended to include the net new functionality? What preceding inputs are needed for this new scenario and how is the output consumed? The simplicity of having the landscape in place along with management's discipline to be lean about the processes they adopted created a nimble CRM solution for the business that could expand as the business grows. It also matured everyone's expectations about how process work should function and reduced the noise and confusion that had been in place prior to implementing this basic framework.

Each question can be viewed as a way to explore impact on the greater process ecosystem and drive the business stakeholders' awareness. In this way, the business value is understood, and employees understand their contribution and impact as process operators. This synchronicity drives adherence to existing processes, efficient analysis of proposed changes, and transparency within the organization and, where appropriate, externally. The transition from one way of operating to another can be visually represented in process maps and related materials and can shorten the changeover timeline significantly.

The Result

Using this approach, the transition from the legacy CRM application to the new one took three and a half months. Without this approach, the project might have taken a year or more to implement. The president of the company remarked a week after the new CRM was operational that he had never seen a project be so successfully delivered in such a short period of time. Talented project team members aside, adoption of the process-oriented architecture foundation was the key to success because it was possible to deliver the solution with high quality and within a tight timeframe.

Case Summary

We recommend that processes be designed to be adaptive in order to ensure their longevity, low maintenance, and ease of adoption. Having said that, sometimes even well-designed processes break down and need maintenance or overhaul. Typically this occurs when an external condition changes. For example, mergers, takeovers, and major technology shifts that require integration with systems and impact extensive processes are not easily taken into account at the time the process in question is created. Nevertheless, these events happen from time to time and can create a profound disconnect in the process landscape and can have significant ripple effects that need to be handled to avoid systemic failure and profitability problems. In these situations the importance of having the base level of a process architecture and governance in place can speed up assessment of impact dramatically and ultimately provide a list of items that can be used to prioritize actions.

Remove Roadblocks

The Challenge

Catching opportunities to make improvements doesn't stop at work; you can find them sitting in most any waiting room. A favorite lesson from reading Rummler–Brache's *Improving Performance: How to Manage the White Space on the Organization Chart* is that it is critically important to put in place measures that relate to what happens in between processes in addition to what happens within processes. After reviewing thousands of processes maps in my career, I can count on one hand the number of times I've seen process measures appear. In these situations I have coached process designers on the importance of using measures as a key driver to achieve process improvement. Having metrics about end-to-end process performance, cycle time, volume, defect rate, and similar elements is of course important in order to have some understanding of the process performance health. However, measuring independent processes and totaling their performance is not an accurate way to characterize the overall experience that a customer might have. This case example is about hospital emergency room wait times. In Canada, the statistics on the number of hours a walk-in patient with a nonlife-threatening issue waits are dismal. Think in terms of multiple hours; that's right, it is not uncommon to wait more than 10 hours to be seen by a physician. The hospital administration may measure process performance based on the time a patient first sits in the triage chair to the time the physician assesses the patient. As long as that target time doesn't exceed their target, they have a successful process.

The Solution

As a patient, however, you may measure the experience differently. I would simply have a personal measure that calculates the time from initial experience of pain/issue to pain/issue relief. My measure does not divide up every stage of the process for which time can be measured, that is, time waiting for initial hospital emergency room check-in, for triage, for physician assessment, for discharge, for drive time to the pharmacy, for the time for the prescription to be filled, and finally for time before medication has a satisfactory result. The hospital's measure misses the periods of time in between discrete processes. For example, if I needed an X-ray, the time between the instruction to get an X-ray and actually receiving an X-ray is discarded. The hospital views it as a net new unrelated process. As a patient in pain, I doubt you would agree. The solution is to emphasize the customer experience for those who benefit from the process's purpose—in this case, the patient. By aggregating the discrete processes into a superprocess, measures related to mean time to resolution can be targeted in order to ensure that any routine set of superprocesses operate using the leanest method possible. How can this be achieved in this example?

The Result

Having the proper medical equipment on hand for the treating physician to use during the initial assessment may be one solution. If an X-ray is required, the transport time to the X-ray room, the wait time for the X-ray technician, and the transport time back to the assessing doctor who is waiting for X-ray results could be eliminated. This could shave hours off the overall resolution time for this type of issue.

Case Summary

While annoyed patients may not be enough of an incentive to address the long wait time issue, hospitals are often driven by the financial impact of long wait times. Having measures in place that demonstrate the increased patient intake and the resultant positive cash flow can be a powerful message. Finding waste in processes reduces labor costs that can be turned into higher profit and/or investment in expanded services and staff. A critical eye on the end-to-end process can locate these cost holes more easily. This is done by using a combination of process ecosystem mapping techniques and associated performance measures. By using alternate-path process scenarios, an improved outcome can be modeled and later tested in a sample set of hospitals. Based on the outcome, the process can be further improved and/or released for use by a larger number of hospitals. Even small process improvements can translate into many millions of dollars of savings and justify the focus on process work.

Reward and Celebrate

The Challenge

I hate being on hold. It got me thinking about why companies allow this to happen; after all, these customers are potentially going to spend money. Walking the floor of a call center operation, you start to understand how it all works and how little understanding there is for the customer experience. Call centers notoriously underpay their workforce and enforce severe consequences for employees who don't handle the targeted number of calls during a shift. Employee turnover is usually high as either the workers burn out from stress or management isn't getting the right level of performance out of their workers to turn enough profit. However, there are usually a few high performers who are able to take the stress out of their job, meet their targets, and manage to stick around as long-term employees. Every call center wants to keep these workers, as they seem to have broken the secret code to success. In fact, businesses want to have more of these workers and fewer of the ones who don't seem to keep up.

The Solution

Finding the right balance between encouragement and discipline in order to get employees to follow the planned course of action can be a real challenge. Reinforcement of compliance through recognition and reward helps to establish expectations and drive a culture of improvement for everyone. An environment that encourages employees to spot problems and report them so that they can be resolved has proven to have more successful outcomes and does so with a happier workforce. Understanding the purpose of a process and having an understanding of its goals, procedures, and expected results creates the opportunity for all employees to operate the process effectively. Setting consequences—positive ones such as a financial reward, prize, or even simple acknowledgments for doing the job well—go a long way. Other consequences, the ones we like to discourage, are equally important when building a culture of improvement.

The Result

Transparency of process adherence is important in achieving awareness of potential problem spots in process design, as well as human operator contribution. Call centers typically use large displays of metrics to show performance of their employees. Figure 9-1 depicts an example of agent service performance.

These displays inform those who see the results and drive specific action. When there is a clear advancement in performance, call center management can reward their top performers with high praise and some sort of financial incentive to continue this performance. The undercurrent to this high praise encourages all other employees to follow the successful pattern and process used by the top performer. Similarly, management may also identify employees on a *wall of shame*, which identifies specific people and their low performance results in an attempt to drive improvement and discourage more low performance. Poor performance can be caused by many things. However, in this example, a process expert would focus on the specific processes to determine how the as-is process is contributing to one set of performers doing well and another set doing poorly. Based on these findings, the expert may model process changes that remove variability in performance and test the new process to see if all performers can follow an improved process that drives high performance.

Another example examines how recognition can drive results. Let's look again at the example of implementing the new CRM solution. After launching the new tool, the top sales executive reviewed the *neglected accounts* report and noticed that one sales person was servicing several of these neglected accounts. Within the application was a message board in which the executive asked the system administrator how he could attach a comment in order to ask the sales person to explain why he had so many neglected accounts. The executive didn't realize that her inquiry would be visible to all users in the CRM application. Her intent was to target the communication to the individual sales person in order to drive

Average Handle Time

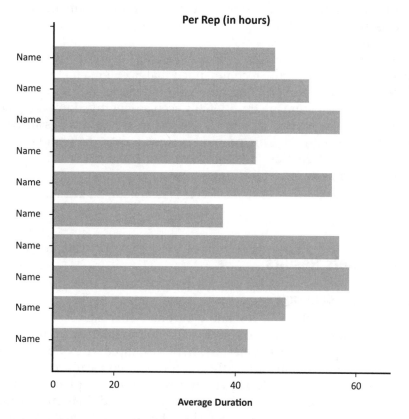

This is the average duation of all cases per rep for the current Fiscal Quarter

FIGURE 9-1 Call Center Performance

action and to increase revenue for the company. This quest for individual accountability may have driven one of the single biggest success factors for future recognition. Within a few hours, the sales person contacted the neglected accounts, and there was an added benefit. Other accounts that were also on the list associated with other sales people were suddenly moved off the report, too. The entire sales organization had received the message and realized that the executive was watching over their performance with a critical eye, and that drove improved behavior and action. The process of servicing accounts was well defined. However, as we have been seen often in this text, attention to process performance through things like the neglected-accounts report helps shape action and bring recognition and reward.

Case Summary

Ensuring the right outcome is something of a management art. The measurements put in place at strategic points in active processes should radiate information about the performance of the process. Business intelligence information is imperative to fixing broken processes, as well as to fine-tuning processes that perform well. It is highly recommended that Six Sigma methods be used to analyze process performance and pinpoint bottlenecks and trouble spots. Analysis projects of this nature are traditionally seen as lengthy; however, when an established process-oriented architecture model is used, the assessment time can be significantly reduced. This provides valuable data that can be used to model improvement scenarios to select from and understand the downstream and upstream impacts on the business landscape.

---

30 Days Gets Results

The Challenge

You get the call. You know the one: "This is CxO so and so, and I need to know if we are meeting with our regulatory compliance regulations asap."Where do you start? In this example we explore the exact situation. It occurred in early 2012 when I received a call from an executive about our company storing credit card numbers in plain text. Customers who placed orders with the company provided their credit card information via phone, email, or website. The credit information was electronically stored to help in processing customers' order, either systematically or by the employees who handled the customer calls. The process had been in place for more than a decade before the company recognized that it was out of compliance with payment card industry (PCI) regulations. The regulation stipulated that credit data could not be displayed in plain text for employees/anyone to view in order to prevent identity theft and fraud. Because there was an unknown number of systems and places within applications that credit data could be stored, we knew we had a problem.

The Solution

Faced with a significant penalty and recognizing that the risk of identity theft had been unmanaged for more than a decade, the executive demanded that all credit data be masked to ensure no employees could see customer credit card information. The seemingly simple request created a firestorm of activity and meetings. Leaders within the order-processing business unit were familiar with the practice of storing credit card information in plain text and knew that credit card numbers were readable by hundreds of people who had access to the order-processing system. A team attempted to isolate all mechanisms that would need to be changed in order to encrypt the credit card information. However, there was confusion about just how many places the credit information could be, and that made senior leadership concerned about how long the company had been at risk of losing its customers' credit information. Days went by. What first appeared to be complete identification of all the places that credit data appeared turned out to be inaccurate. The team was back in a room trying to calm the senior leadership's concern by claiming that because the company had never had an incident before, their concern wasn't warranted. The division director knew that this information was stored in dozens of places, and she reached out to the process team for help to fix the gap for good.

The Result

This was the point where the process-oriented architecture model was put in place and a solution to the problem found. In short order, the team huddled together and mapped the company's entire order-to-cash process. There was a tremendous focus on getting everything mapped, placing the business rules and conditions out in the open, and achieving clarity on misunderstood ways in which the business had been running. Special care was taken to identify all the customer credit data and the systems that were part of the interactions. The team identified all sensitive data within one day. In doing so, the team initiated a segment of a larger process ecosystem that, over the coming months, enabled the organization to spot bottlenecks and gaps in process coverage and control and to speed up delivery simply by enabling the organization to see how things were working. The credit data were programmatically masked so that the company would be compliant with PCI legislation, and all customer information was safe and secure.

Case Summary

Not all organizations have their process landscape documented; in fact, most do not. What seems like a daunting task might actually turn out to be the single biggest competitive differentiator for a successful business. Those who recognize the need for architecting their processes are more likely to have the end result for their customer be successful and profitable. Using the process-oriented architecture framework, we suggest that you take on the toughest process issue first in order to get a baseline of the business engine. It is through steady evolution that the process landscape not only emerges but also becomes a valuable tool.

In one of our most successful process transformation experiences, we plastered the walls of the head office hallway with end-to-end process maps for all to see. This generated self-correction of the maps and corresponding adjustments through governance monitoring. In doing so, the organization became abuzz with excitement about improvement opportunities, and we found ourselves engaged in high-value discussion with reenergized employees about how they could make a difference. The result was improved employee satisfaction scores and creation of the momentum necessary to achieve a culture of measurement and improvement.

Chapter Summary

In this chapter, we reviewed Case examples to illustrate the value of the concepts presented in this book. Following is a summary of these examples:

- *Recognize What Failure Is:* The Amazon.com experience demonstrated the importance of consistency in experience through process continuity and quality.

- *Achieve Customer Delight:* The winter coat story revealed the expectations that can be set and met if well thought out.

- *What Can Go Wrong, How To Make IT Right:* This example reviewed process design with flexibility in mind.

- *Remove Roadblocks:* Here we identified how measures in process outcomes can trick organizations into misunderstanding the service impact to a customer by not measuring the process experience end-to-end from the customer's perspective.

- *Reward and Celebrate:* This example retold the success achieved by using incentives to drive process maturity.

- *30 Days Gets Results:* Here we noted how process maps and associated attributes can provide strategic value and speed to an enterprise.

Chapter Preview

To this point, we have covered principles and practical examples in which one can understand the usefulness of *The Process Improvement Handbook*. In the next chapter, templates and their associated instructions are presented as companion guides to help an organization build their own tool kit toward Process Improvement.

CHAPTER 10

Process Improvement
Templates and Instructions

T oday, business operations and project efforts often create, use and store common documents. These documents often have to follow specific standards and be in specific formats while also containing content that is important and relevant to stakeholders and readers. This task is often made easy by the creation and use of document templates. A template is a master copy of a publication used as a starting point to conduct various business efforts including project management, process improvement, and various other activities. It may be as simple as a blank document in the desired size and orientation or as elaborate as a nearly complete design with placeholder text, fonts, and graphics that need only a small amount of customization. A template can be used to jumpstart the design process or as a way to help insure that individuals in an organization create memos, processes, project artifacts and other documents in a uniform fashion. This chapter provides examples and guidance embedded within templates that Process Improvement practitioners and organizations can use to build and deliver consistent and sustainable service to customers.

NOTE *The templates provided in this chapter deliver various examples and instructional material for readers and practitioners to use as they employ these techniques within their organizations or environments. Please visit mhprofessional.com/pihandbook to download customizeable versions of these templates.*

5S Template

This template is used to help achieve organization, standardization, and cleanliness in the workplace. It is used to reduce waste and optimize productivity by maintaining an orderly workplace. Implementation of the 5S method helps organizations organize themselves; and is typically one of the first Process Improvement techniques that organizations implement (see Figure 10-1). A well-organized workplace can result in more efficient, safer, and more productive operations and can boost the morale of an organization's workforce. The 5S pillars or practices are based on five Japanese words:

Figure 10-1 5S Template

Number	Item	Sort	Score 0 1 2 3 4	Set in Order	Score 0 1 2 3 4	Shine	Score 0 1 2 3 4	Standardize	Score 0 1 2 3 4	Sustain	Score 0 1 2 3 4
1		Are items properly organized and stored?		Are items not needed for the current job assignment put away?		Are cleaning supplies easy to access?		Have specific job tasked been assigned?		Does the workforce reflect on opportunities for improvement as part of their assignment?	
2		<sort criteria>		<set in order criteria>		<shine criteria>		<standardize criteria>		<sustain criteria>	
3		<sort criteria>		<set in order criteria>		<shine criteria>		<standardize criteria>		<sustain criteria>	
4		<sort criteria>		<set in order criteria>		<shine criteria>		<standardize criteria>		<sustain criteria>	
5		<sort criteria>		<set in order criteria>		<shine criteria>		<standardize criteria>		<sustain criteria>	
6		<sort criteria>		<set in order criteria>		<shine criteria>		<standardize criteria>		<sustain criteria>	
7		<sort criteria>		<set in order criteria>		<shine criteria>		<standardize criteria>		<sustain criteria>	
8		<sort criteria>		<set in order criteria>		<shine criteria>		<standardize criteria>		<sustain criteria>	
9		<sort criteria>		<set in order criteria>		<shine criteria>		<standardize criteria>		<sustain criteria>	
10		<sort criteria>		<set in order criteria>		<shine criteria>		<standardize criteria>		<sustain criteria>	
Section Scores			/40		/40		/40		/40		/40

Grand Total /40

5S Score /200

Instructions:
1: Add each column to get a score out of 40
2: Add columns to get a score out of 200
3: Divide score by 200
4. Multiply by 5 to get 5S score

Score	Condition Observed	Score 0 1 2 3 4
4	Very Good	
3	Good	
2	Ok	
1	Bad	
0	Very Bad	

FIGURE 10-1 5S Template

Sort (Seiri), Set in Order (Seiton), Shine (Seiso), Standardize (Seiketsu), and Sustain (Shitsuke). These pillars encourage workers to improve their working conditions and then reduce waste, eliminate unplanned downtime, and improve processes.

You can customize the criteria in each row to suit your business situation. Use the 5S template to inspect and evaluate the specific score based on the following 5S criteria to determine the total 5S score on a regular cadence while helping the organization make appropriate changes that drive improved scores and outcomes.

5S Elements:

1. Sort: Eliminate all unnecessary tools, parts, and instructions keeping only essential items and eliminating what is not required from all work areas.

2. Set in Order: Arrange all tools, parts, supplies, equipment, manuals and instructions in a way that the most frequently used items are the easiest and quickest to locate in order to eliminate time wasted in obtaining the necessary items for use in a process or operation.

3. Shine: Improve the appearance of all work areas, and conduct preventive housekeeping which is the act of keeping work areas from getting dirty, rather than just cleaning them up after they become dirty. Clean the workspace and all equipment, and keep it clean, tidy and organized ensuring everyone knows what goes where and everything is where it belongs.

4. Standardize: All employees in an organization should be involved with creating best practices and implementing those best practices the same way, everywhere, and at all times. Any processes, procedures, work areas and other settings should be standardized wherever possible so that all employees doing the same job are working with the same tools.

5. Sustain: Indoctrinate 5S into the culture of the organization and maintain and review standards. Maintain a focus on this new way of operating and do not allow a gradual decline back to the old ways. Implement continuous improvement into general practices and operations to ensure the organization is continually making itself better.

5 Whys Template

This template is used to generate root causality and expose insight into the problem definition. The 5 Whys method helps to determine the connection between symptoms and true root cause. As a simple method, it simply probes deeper into a situation or problem until enough insight can be revealed to determine the originating event causing the problem.

<Project Name>	
Problem Statement: <Describe the problem or the current state, as it is presently understood. Then, answer the 5 Why's section to help identify the true root cause of the problem. Five iterations of questions is a recommendation, however, more could be needed until the true root cause is identified.>	
Why Question	**Answer: Because...**
1. Why do we have an increase in customer complaints?	• Because of late orders
2. Why do we have late orders?	• Because the orders are not complete when received by Order Processing
3. Why are orders not complete when received by Order Processing?	• Because the customer didn't provide all the necessary information
4. Why didn't the customer provide all the necessary information?	• Because the form used isn't up to date with all the needed fields of information
5. Why doesn't the form capture all the needed information?	• Because the form did not get updated when the order processing process changed last quarter which changed what fields of information was needed
Suspected Root Cause: <A concise description of the suspected root cause after completion of the 5 Why's exercise>	
Example: The increase in customer complaints due to late order arrival is due to missing required information at the time of order placement. This problem can be remedied by updating the order form, and ensuring that when/if the order process changes that the order form is reviewed to ensure it continues to capture appropriate information.	

FIGURE 10-2   5 Whys Template

The template provided can be used to capture a specific problem as reported and the corresponding natural questions and their response. The template can be customized by any number of Why questions as needed to determine root cause. Another method of the 5 Whys is shown in Figure 8-1 (the Ishikawa Diagram), often referred to as the fishbone diagram. The Ishikawa method also uses the Why model, starting from the tail of the fish diagram and working toward the head until the root cause is identified. These cause-and-effect methods are very powerful tools that can be employed in both complex and simple improvement projects. Figure 10-2 outlines a standard 5 Whys template.

A3 Template

This template is used to solve problems, report project status, and provide visibility to the to-be state. Created by the Toyota Motor Company, the A3 has become a best practice format for presenting concise information that results from structured thinking about a problem. The name A3 is derived from the size of paper the information is presented on; A3 for metric and 11×17 in imperial measurement. A3 in the process industry has become the gold standard in Lean Six Sigma and other methodologies for presenting problems and solutions, as well as the problem-solving facts that need to be understood by less-involved audiences. The A3 can be easily mastered in less than 10 minutes. When filled out completely, the A3 can become a powerful aid in bringing visibility to a given situation; in practice this leads to higher degrees of focus and support for solution proposals (see Figure 10-3 for a standard A3 template). The A3 template is particularly helpful when presenting complex problems and information to executives.

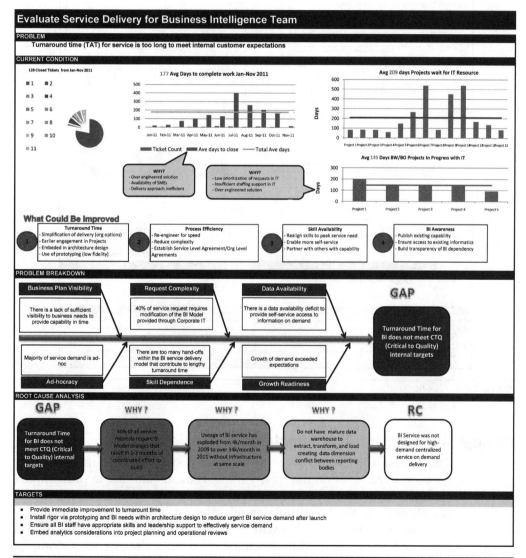

FIGURE 10-3 A3 Template

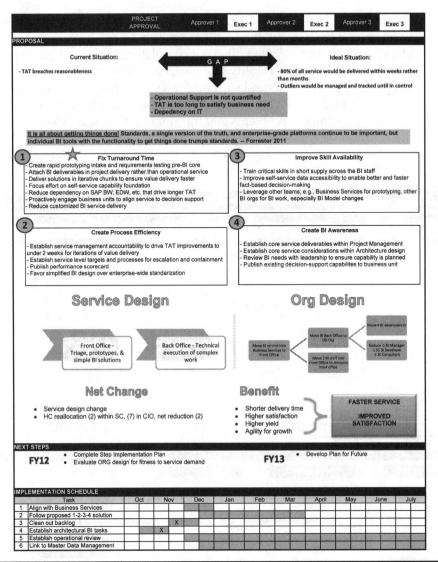

Figure 10-3 *(Continued)*

Change Management Plan Template

This template is used to organize the adoption of change within an organization in recognition that many improvements fail due to professional change management practices. Many process improvement projects fail simply because there was not enough attention given to change management as a core responsibility within the project. The template provided is one tool that can be used to articulate the overall strategy for staying on top of change management (see Figure 10-4). The change management

Change Information	
Change Title	\<Use a meaningful title that can be understood at a glance>
Change Implementation Date	\<Date that the change will be realized>
Change Management Board Review Date	\<Date of the review board>
Change Requested On	\<Date the change was requested on>
Change Requested By	\<Name of accountable person requesting change>

Goals & Objective

\<A successful improvement project can only be effective when it is accepted and appropriately implemented by the affected department. The change management plan is focused on increasing the 'acceptance' of the improvement project.

$$Quality \times Acceptance = Effectiveness$$

Reiterate the goals and business case from the Project Charter is a concise manner to support the change implementation.
* Explicitly why the change is being implemented; is it to address opportunities or overcome challenges?
* The value-add and rationale of the proposed change?
* Defines and clarifies the goals that the change is focused on achieving.>

Description of the Change

\<Provide a description of the current state, future state and the change that is required to bridge the gap>

Personnel Implications	
Group/Individual	**Impact**
\<Identify groups/individuals that will experience a change>	< Describe how will the group/individual be impacted by the implementation of the changes>

Process Implications

\<Provide a list of processes that will be impacted and describe the change to each process; recommend using the Process Landscape to focus awareness on impact areas>

Technology Implications

\<Describe the impact on each system that will be changed>

Note: Refer to Project Plan for any Action Plan tasks and the designated Project Communication Plan for project and implementation related communications

FIGURE 10-4   Change Management Plan Template

plan can be used to help organizations identify the intended change and the impacts that are anticipated, describe the future state vision, address critical elements of the change, and reveal the reasons for the change as well as the benefits to both the company and the employee population. Through a series of questions that strive to tease out the salient points of the change, the template can be tailored based on the business situation to target communication messages (see Communication Plan Template).

Check Sheet

A Check Sheet is used to collect and structure data so that decisions can be made based on facts, rather than anecdotal evidence. Data are collected and ordered in a table or form by adding tally or check marks against problem categories. It allows a group to record data from either historical sources or as activities are happening so that performance patterns or trends can be detected. A Check Sheet presents information in an efficient, graphical format that registers how often different problems occur and the frequency of incidents that are believed to cause problems (see Figure 10-5). Check Sheets focus on an event or problem that can be observed and when the data will be collected, for how long, and from what sources. Each time the targeted event or problem occurs, the template is used to record the data in a consistent and accurate manner so that data collection is simple and aids in identifying and understanding performance patterns and trends.

Check Sheets are primarily used to:

- Identify what is being observed in a particular process or environment

<Document Title>		
Number	Description	Frequency
	Identify the type of item (error, problem, etc.)	For each occurrence of the item add to the count to indicate the number of times the item happens
1	Problem 1	II
2	Problem 2	I
3	Problem 3	IIII
4	Problem 4	III
5	Problem 5	IIIII IIIII II
6	Problem 6	II
7	Problem 7	I
8	Problem 8	
9	Problem 9	III
10	Problem 10	IIIII IIIII IIIII IIIII

FIGURE 10-5 Check Sheet Template

Audience	What	When	Method	Owner	Required Approval	Feedback Loop
• Include the project audience(s) • Include internal and external stakeholder groups	• Determine what groups need to know and expect • Communicate only information • Identify key project messages that need to be delivered on a recurring basis	• Define when communications will be provided	• Identify the best delivery vehicle for communication face to face; meetings; presentations; email	• Assign responsibilities for crafting and delivering communications listed	• Document the approval process required	• Create feedback loops to help evaluate the communications

FIGURE 10-6   Communication Plan Template

- Collect data in a easy and simple fashion
- Group data in a way that makes the data valuable and reliable to stakeholders

Communication Plan

A Communication Plan is used to ensure awareness and transparency throughout the lifecycle of the improvement project and to drive positive change acceptance. The communication plan is a strategic tool that identifies all the stakeholders, the types of communications they each need, and the frequency in which the communications occur. The communication plan template can be used to provide a cadence to communication launches and as an aid in structuring approvals as well as what messages are sent and received at various intervals. Figure 10-6 shows a standard Communication Plan Template. The communication plan is a road map that is essential in all process improvement initiatives. It guides messaging in a meaningful and professional fashion that is consistent with the change management discipline needed by an organization to absorb improvements at the various stages of the project. The communication plan helps the entire organization keep pace with the direction of the improvement.

Dashboard Measurement Plan

A Dashboard Measurement Plan is used to identify the inputs, outputs, measures, and mechanisms for monitoring performance. The template (Figure 10-7) provided organizes the measurements needed by business goals and objectives and type of measure required to provide the measure. Effective dashboards ensure that accurate information is provided and that effective action can be taken. Information flow that is close to real-time is

Unit Being Processed	Quality or Process Indicator	Measurement	Display	Input Measure (√)	Process Measure (√)	Output Measure (√)
• The object that is passing through and being changed in the process	• Process Indicator – used to understand how the process and inputs are performing • Quality Indicator – used to understand how well we are meeting customer requirements	• Control Chart; Run Chart; Histograms; Pareto Charts	• How do you want the information displayed for the measures? • To what level do you want to drill down in the information?	• Measures that are placed on the suppliers to the process to ensure quality inputs are received	• Measures that re placed on your internal processes • Measures for internal customers or those that influence output measures	• Measures used to determine how well customer needs and requirements are met
Insurance Claim	Invalid Claims	Invalid Claims as a percentage of total claims versus time of receipt	Run Chart		√	

FIGURE 10-7   Dashboard Management Plan Template

presented in dashboards so that it can be used in time-sensitive response situations, whereas more traditional reporting analytics can be used in situations where real-time is not needed or practical. Regardless of the frequency, the information presented must address the core business objective in order to be useful. The dashboard measurement plan challenges the thinking about what measures are needed in the context of what action will follow on as a result of having the information in hand. In this context, the template should be populated with targeted and well-described measures that do not overlap with other measures.

Data Collection Plan

A Data Collection Plan is used to identify the sources of and the types of data used for analysis and ongoing monitoring of performance. The data collection plan template should be used to ensure improvement solutions are solidly built on reliable data (see Figure 10-8). Six Sigma experts and

Measure	Type of Measure	Type of Data	Operational Definition	Sample	Display
Specify the name of what is to be measured (e.g. billing cycle time)	Specify if it is an output, input or process measure	Discrete or Continuous	Define what is being measured, how it will be measured and recorded	- Number to be sampled - Where in the process is the sample - Frequency of sample	What tools will you use to display the data (histogram, run chart, control chart etc)
Cycle Time	Process Measure	Continuous	The time from receiving request to completing request and notification back to the requestor.	100% of requests will be measured continuously	Control Chart

FIGURE 10-8   Data Collection Plan Template

statisticians rely on valid data in order to make decisions and infer improvements. Consequently, having a credible data collection plan is foundational. The collection plan honors the integrity of the data-gathering process and resultant data in order to ensure no bias enters the process. The plan addresses predata collection, during collection, and after collection as well as why the data are collected.

Disconnect List Template

A Disconnect List Template identifies the gaps in a given process or set of processes. The template (Figure 10-9) is used to uniquely itemize all disconnects and is used in conjunction with a process map to ensure reference linkage between the artifacts. Attributes that can be customized are used to help assess the level of importance against a specific goal (e.g., customer impact, revenue impact, legal exposure) and impact damage that is occurring or will occur if left as is (e.g., increased cost, damage to brand reputation, increased risk to human life). The Rummler–Brache methodology also encourages itemizing perceived disconnects, as they also must be addressed in an improvement project. The disconnect list can be used to centrally prioritize and sequence solution attention and drive responsive action accordingly. The list can be used as a quality checklist for testing proposed solutions to ensure the solution resolves all known disconnects and does not introduce new disconnects.

Force Field Analysis

Force Field Analysis is an effective team-building tool and is especially useful when it is necessary to overcome resistance to change (see Figure 10-10 for a Force Field Analysis Template). When a change is planned, Force Field Analysis helps organizations see what the effects will be on the larger business landscape by analyzing all of the forces impacting the change and weighing its pros and cons. Once these forces have been identified and analyzed, it is possible for an organization to determine if a proposed change or improvement is viable. There are several factors that can be analyzed, including people, resources, attitudes, traditions, regulations, values, needs, and desires. Force Field Analysis helps identify those factors that must be addressed and monitored if change is to be successful. Forces that help an organization achieve the change are called driving forces, and any force that might work against the change is called a restraining forces.

No.	Priority I=High M=Med L=Low	Type O=Org P=Proc J=Job	Category	Full	Addresses Part	Is Step No.	Disconnect Description	Disconnect Impact	Additional Information to be Gathered	Source of Information	Quick Win Possibility? Yes/No	Technology Required Yes/No
1	H	P	Resources	Yes		11.9	There are too few computers available to handle the number of users during working hours	Staff time is being wasted as people wait in line to use a computer. This is impacting the number of widgets produced per day negatively.	There is an unknown number of widgets that cannot be produced that is contributing to lost revenue.	Users of the computer lab provided reports of the problem, and was confirmed by observing the situation over a period of two weeks	Yes	Yes
2												
3												
4												
5												
6												
7												
8												
9												

FIGURE 10-9 Disconnect List Template

290

Driving Forces (+)   Resisting Change (–)

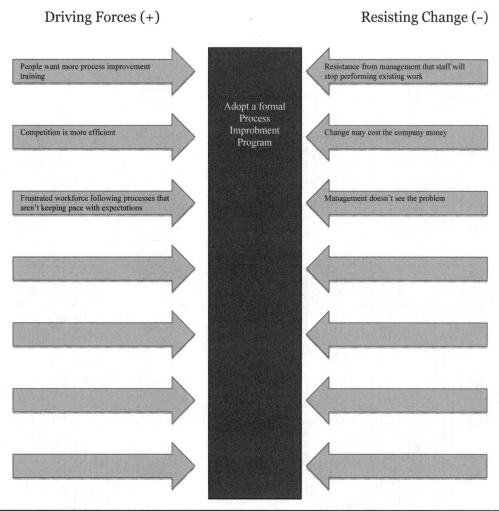

People want more process improvement training

Competition is more efficient

Frustrated workforce following processes that aren't keeping pace with expectations

Adopt a formal Process Improbment Program

Resistance from management that staff will stop performing existing work

Change may cost the company money

Management doesn't see the problem

FIGURE 10-10   Force Field Analysis Template

Hoshin Kanri

Hoshin Kanri articulates the strategic goals, strategies, initiatives, account-abilities, and measures of an organization on a single page. The template demonstrates the linkage between these elements and is intended to indi-cate the correlation relationship (as shown in Figure 10-11). The template should be used to incite clarity of purpose for organizations involved in improvement efforts. The Hoshin Kanri model is simple and effective for prioritizing process improvement initiatives. Each segment of the Hoshin Kanri builds on the foundation of the strategic goals of the company while maintaining awareness of each foundational element that proceeds it. The Hoshin Kanri model also incorporates the key measures and personnel accountable for meeting the objectives identified. Process improvement

Global Information Technology Organization - Hoshin Kanri Model

Legend: 2 Direct Effect 1 Indirect Effect

Accountability — Functional Leads:
- VP
- Director IT
- Director ERP
- MGR Service Management
- MGR Application Support
- MGRs Infrastructure
- MGR External Support

Contribution — Improvement Targets:
- Have strategic plan
- Improve Operation
- >10% reduction in cost/elimination of waste
- >20% of effort spent in new FY13 value-added activities
- <25% of time required to complete manual activity
- >50% of process standardized
- 50% reduction in employee service related cycle time
- >75% of work standardized
- Improve Employee Satisfaction
- Work Alignment to Strategic Goals
- Provide Operational Stability
- Alignment of Projects to Customer Needs

Level 1 - Improvement Priorities:
- Make infrastructure secure
- Make infrastructure reliable
- Enable 'work anywhere' capability
- Make infrastructure fast
- Be efficient
- Improve everything

Level 2 - Improvement Priorities:

Improve Operational Stability
- Strategic Initiatives
- Strategic Initiatives
- Strategic Initiatives
- Strategic Initiatives
- Strategic Initiatives
- Strengthen Information Security

Ongoing modernization of enterprise IT architecture and policy
- Establish Virtualization
- Expand Virtualization to critical services
- Establish Enterprise Architecture across IT
- Establish ITIL Framework service management
- Create Private Infrastructure-As-A-Service (IaaS)
- Refresh aged computing equipment

Strategic Initiatives
- Create new data center in X location
- Replace iPhones with BlackBerry 10 infrastructure
- Support New Product Introduction
- Launch new project for X

FY13 Annual Objectives:
- Make a lot of money
- Be lean
- Make Customers love us
- Be better than competition
- Innovate
- Be everywhere we need to be
- Balance future needs with operational needs
- Save money
- Have high quality
- Stay true to the company mission and values

Correlation

FIGURE 10-11 Hoshin Kanri Template

292

projects benefit from using this artifact to ensure their improvement falls in line with the corporate strategy and/or addresses disconnects in alignment early in the project.

Lessons Learned Survey Content

Lessons Learned Survey Content can be used to gather lessons learned about an improvement initiative with a view toward future improvement. The template strives to capture the raw feedback from participants (see Figure 10-12). Questions presented can be tailored to suit the needs of the situation. In general, the survey should elicit both things that went well and areas for improvement. Surveys can be completed in person, electronically, or anonymously depending on the comfort of the recipient. It is important to solicit feedback using surveys at varying intervals during the improvement initiative in order to adjust strategies such as change management plans, communication plans, or solution designs based on the feedback received. Lessons learned should be analyzed, distilled into a master set of lessons learned, and redistributed to recipients in order to acknowledge that their feedback was captured and understood. Further, the lessons should be distributed to other improvement project teams to serve as early input that can contribute to more positive results. All lesson learned information should be stored in an appropriate repository for ease of access.

Introduction

This survey has been created to solicit feedback on the <project name> and to identify how the project was conducted. This survey is meant to capture lessons learned from the project while they are fresh in people's mind. This survey focuses on the actions – not the people – to identify what worked and identify areas for improvement. The results of this survey will remain anonymous.

Please complete this survey and e-mail the results, as an attachment, to <insert e-mail address> by <insert date>.

The survey results will be compiled (anonymously) and discussed in detail at an upcoming Lessons Learned meeting. The feedback received will help point to particular area that needs extra exploration in the group meeting and identify potential areas for improvement.

The results from the survey and Lessons Learned meeting will be summarized in a Lessons Learned document and recommendations will be passed to future teams.

To use the form:
- To 'check' a box, double click on your selection
- Under 'default value', click on 'checked'

Figure 10-12   Lessons Learned Survey Content Template

General Project Issues and Communication

1. Are you satisfied with the finished deliverable?

☐ Very   ☐ Somewhat   ☐ Not Very   ☐ Not at all

If satisfied, what was good about it?

If not, what was wrong with it?

2. Were the project team meetings were efficient and effective?

☐ Very   ☐ Somewhat   ☐ Not Very   ☐ Not at all

If yes, what went well?

If no, what could have been done differently?

3. Was the entire team committed to the project schedule?

☐ Very   ☐ Somewhat   ☐ Not Very   ☐ Not at all

If not, what could have been done differently?

4. How involved did you feel in project decisions?

☐ Very   ☐ Somewhat   ☐ Not Very   ☐ Not at all

If you did not feel involved, what decisions did you feel left out of?

5. How efficient and effective was communication between the project sponsor, project manager and team members?

☐ Very   ☐ Somewhat   ☐ Not Very   ☐ Not at all

If efficient and effective, what went well?

If not, what could have been done differently?

Solution Development Life Cycle (Define, Analyze, Design, Develop, Test, Deploy)

Define

1. How clearly defined were the objectives for this project?

☐ Very   ☐ Somewhat   ☐ Not Very   ☐ Not at all

If clearly defined, what went well?

If not, what could have been done differently?

2. How clear were you on your role in the project?

☐ Very   ☐ Somewhat   ☐ Not Very   ☐ Not at all

If clear, what went well?

If your role was not clear, what could have been done differently?

FIGURE **10-12**   *(Continued)*

<u>Analyze</u>

3. How effective was the requirements identification process?

☐ Very   ☐ Somewhat   ☐ Not Very   ☐ Not at all

If effective, what went well?

If not effective, what could have been done differently?

<u>Design</u>

4. How effective was the design (or implementation specs)?

☐ Very   ☐ Somewhat   ☐ Not Very   ☐ Not at all

If effective, what went well?

If not, what could have been done differently?

5. How effective were the design reviews?

☐ Very   ☐ Somewhat   ☐ Not Very   ☐ Not at all

If effective, what went well?

If not, what could have been done differently?

6. How effective were the functional specs?

☐ Very   ☐ Somewhat   ☐ Not Very   ☐ Not at all

If effective, what went well?

If not effective, what could have been done differently?

<u>Develop</u>

7. How effective was the test plan?

☐ Very   ☐ Somewhat   ☐ Not Very   ☐ Not at all

If effective, what went well?

If not effective, what could have been done differently?

FIGURE **10-12**   (*Continued*)

8. How effective was the training plan?

☐ Very   ☐ Somewhat   ☐ Not Very   ☐ Not at all

If effective, what went well?

If not effective, what could have been done differently?

Test

9. How effective was the testing process?

☐ Very   ☐ Somewhat   ☐ Not Very   ☐ Not at all

If effective, what went well?

If not effective, what could have been done differently?

10. How effective was the interaction/cooperation between technical sub-teams?

☐ Very   ☐ Somewhat   ☐ Not Very   ☐ Not at all

If effective, what went well?

If not effective, what could have been done differently?

Deploy

11. How effective was the deployment process?

☐ Very   ☐ Somewhat   ☐ Not Very   ☐ Not at all

If effective, what went well?

If not effective, what could have been done differently?

Open Comments

For the next phase or project, how or what could we improve on the way the project was conducted?

FIGURE 10-12 *(Continued)*

Meeting Minutes Template

A Meeting Minutes Template provides an example of an effective meeting minutes record that captures the essential points of a meeting, including decisions, action items and accountable personnel, timelines, expectations

Minutes	Date & Time	Type (Conference Call/On Site)
MEETING CALLED BY	\<Name of meeting host\>	
TYPE OF MEETING	\<Purpose of the meeting\>	
FACILITATOR	\<Name of facilitator if different than host\>	
ATTENDEES	• \<List of attendees\>	
ABSENT	• \<List of absent attendees\>	
MINUTES	\<Record of decisions, actions, noteworthy items\>	

Agenda topics

DISCUSSION	\<Topic\>	
\<Noteworthy items related to topic\>		
CONCLUSIONS		
\<Decisions, important statements\>		
ACTION ITEMS	PERSON RESPONSIBLE	DEADLINE
\<List all action items\>	\<Identify action owner\>	\<Record due date\>

FIGURE 10-13   Meeting Minutes Template

and points for consideration, attendees and absences, as well as the and place and time of the meeting (as shown in Figure 10-13). After meetings, the minutes should be distributed to ensure all relevant information was captured and is accurate. The minutes formulate a formal record of events and are often referred to for clarification and alignment. Meeting minutes are valuable to process improvement projects as they provide reliable information that is not susceptible to lapses in human memory.

Metrics Chain

A Metrics Chain can be used to ensure traceability between improvement goal and metrics that radiate performance toward that goal. The template uses a simple input/output scheme to derive the relationships and provide insight into the standard of measure used and value realized (see Figure 10-14). Metrics constructed using this template help organize a hierarchy of metrics to avoid duplication and overlap. Additionally, the content can be used to ensure focus is on the right data elements and their use. When starting out, it is recommended that attention be given to the top-level metrics that impact existing performance and determine desired performance. This provides an initial dataset that can be assessed for gaps in metric coverage and then improved. Common advanced metrics include diagnostic metrics around root cause determination, executive insight, and predictive/extrapolator metrics. All of which can be recorded in the metrics chain template.

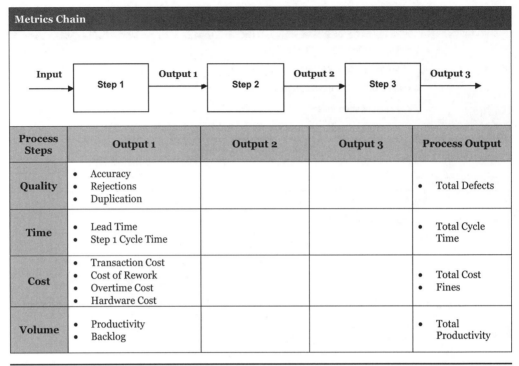

Process Steps	Output 1	Output 2	Output 3	Process Output
Quality	• Accuracy • Rejections • Duplication			• Total Defects
Time	• Lead Time • Step 1 Cycle Time			• Total Cycle Time
Cost	• Transaction Cost • Cost of Rework • Overtime Cost • Hardware Cost			• Total Cost • Fines
Volume	• Productivity • Backlog			• Total Productivity

FIGURE 10-14 Metrics Chain Template

Policy Template

A Policy Template sets the standard for clear organizational policy that ensures consistency and ease of adoption. The template captures the essential elements of policy in the context of how it impacts the organization and related processes (as seen in Figure 10-15). Policy is a foundational consideration in the ecosystem of process design. All processes are expected to adhere to specific policies as they provide rule definition and the framework for compliance. The template provided also captures the attributes associated with the dimension of time, that is, when the policy takes effect, is expected to be reviewed, and, where appropriate, ends. In Process Improvement work policy includes difficult-to-change rules such as governing law and regulatory rules or corporate governance policy at a board level. Policy is holistically apart from the notion of changeable organizational decisions and work instructions that guide how work is accomplished. The latter is often confused as a policy as it can be distributed in a very formal fashion. However, organizational rules of this nature should be assessed to ensure legacy process steps are not being masked as policy. As a general rule, companies have only as many policies as required and many processes that articulate how the company complies with those policies.

Policy Overview	
Policy Title	\<Brief title of the policy\>
Policy #	\<Internal reference number\>
Owner	\<Name of authority for the policy\>
Effective Date	\<Date the policy is in effect from\>
Expiry Date	\<Date the policy expires or needs to be reviewed\>

Policy Description	
Policy Statement	\<Outlining the purpose of the policy and what its desired effect or outcome of the policy should be.\>
Scope Statement	\<Describe why the organization is issuing the policy, whom the policy affects and which actions are impacted by the policy. The applicability and scope may expressly exclude certain people, organizations, or actions from the policy requirements. Applicability and scope is used to focus the policy on only the desired targets, and avoid unintended consequences where possible.\>
Responsibilities	\<A responsibilities section, indicating which parties and organizations are responsible for carrying out individual policy statements. Many policies may require the establishment of some ongoing function or action. For example, a purchasing policy might specify that a purchasing office be created to process purchase requests, and that this office would be responsible for ongoing actions. Responsibilities often include identification of any relevant oversight and/or governance structures.\>

Contacts			
Subject	Contact	Telephone	E-mail Address
\<Policy Clarification\>	\<Responsible Office; Legal\>	\<555-1212\>	\<legal@policy.com\>

Definitions	
Term	Definition
\<sample term\>	These definitions apply to terms used in this document

Related Documents	
Related Processes	\<Identify the processes related to the policy\>
Other Documents	\<Identify any other artifacts; e.g., Legislation, other policies, employment contracts, etc.\>
Forms and Tools	\<Identify other forms and tools related to the policy; e.g., web based signatures of acceptance that the policy is read\>

FIGURE 10-15   Policy Template

Procedure Template

A Procedure Template is used to construct step-by-step instructions that are linked to processes that are governed by policy (as shown in Figure 10-16). The procedure template addresses the purpose for the procedure, the responsibilities and roles associated with operating the procedure, and the specific instructions of the procedure and it references other artifacts and processes connected to the procedure. The template captures procedures in

Procedure Overview	
Procedure Title	<Short name for the procedure>
Procedure #	<Internal unique reference number>
Owner	<Name of organization and individual responsible for the procedure>
Effective Date	<Date the procedure takes effect>
Expiry Date	<Date the procedure expires or is due for review>

Procedure Description	
Purpose Statement	<The brief statement regarding why the procedure exists>
Scope Statement	<Identifies what is in and out of scope of the procedure>
Responsibilities	<Identifies organizations, roles, and individuals involved in the procedure>

Procedures		
Step	Action	Responsible
<#>	<Specific action/instruction needed to complete the stop>	<Role/person>

Definitions	
Term	Definition
<Term used in procedure>	<Articulate the definition of the term in context of the procedure>

Related Documents	
Internal Policies or Documents	<Identify the processes related to the policy>
Other Documents	<Identify any other artifacts; e.g., Legislation, other policies, employment contracts, etc.>
Forms and Tools	<Identify other forms and tools related to the policy; e.g., web based signatures of acceptance that the policy is read>

FIGURE 10-16 Procedure Template

a consistent fashion to aid in adoption speed and familiarity with key reference points within and across multiple procedures. The template encourages simplicity in articulating the procedure to avoid confusion but encourages cross-referencing of other procedures needed for alternate paths within a procedure. All procedures should have a declared owner and monitor who regularly reviews the procedure for accuracy and usefulness. The procedure template should articulate what processes it applies to which supports full traceability and relevance.

Process Map

A Process Map is used to visually describe the flow of activities in a process and outline the sequence and interactions that make up an individual process, from beginning to end. The template outlines the structure in which the activities are placed (see Figure 10-17). It also organizes the

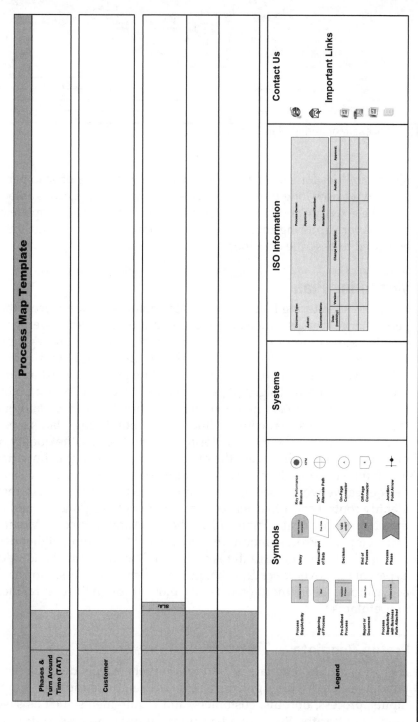

FIGURE 10-17 Process Map Template

Critical To Quality (CTQ)	Process Steps Where the CTQ is Measured	Data Collection Method	Data Collection Frequency	Owner Responsible for Collection
The measurable requirement for CTQ	The step in the process where the CTQ is measured and data collected	Description of the method that will be used to collect the data	The frequency that the data will be collected (hourly, daily, monthly etc.)	The person responsible for the data collection

FIGURE 10-18   Process Monitoring Plan Template

responsibilities for each activity against the dimensions of who performs the activity, how long the activity takes, the controls (e.g., rules associated with the activity), and how the activity is performed. The template offers sections for capturing meta information about the process and the symbols used to describe activities.

Process Monitoring Plan

A Process Monitoring Plan establishes the points within a process that are performance signals to ensure end-to-end measurement coverage. This plan is used to capture the parameters of the agreed-to strategic performance goals and targets, the activities that are in focus that contribute value toward the goal, the frequency of polling performance data, and the specific measures that signal if performance is in control or out of control. Included in the plan are notations about assumptions, risks, and outside factors/conditions that could influence the monitoring plan, as well as who is accountable for the monitoring system. Because the quality of the monitoring plan will impact the quality of the results, it is important to invest time into building out a robust process monitoring plan. Plans are most effective when used as a continuous feedback mechanism that can identify control concerns, help escalate to the right responders, and target the performance issue. Figure 10-18 shows a sample Process Monitoring Plan Template. Process monitoring is essential to building higher degrees of maturity within organizations and should be maintained and updated as part of the governance and change management processes to ensure all active processes are monitored as they move through the natural lifecycle of usefulness.

Process Profile Template

A Process Profile Template is used to comprehensively describe the attributes associated with a given process. The template captures suppliers, inputs, process, outputs, customers, and requirements such as systems. Modeled after the Six Sigma [suppliers, inputs, process, outputs, and customers [SIPOC]) diagram and the more simplified version for capturing similar information found in the IEEE IDEF 0 (Institute of Electrical and Electronics Engineers Integration Definition) specification, the Process Profile Template should be used to expose all factors that influence, control, or are interrelated with a given process (see Figure 10-19).

Process Name <Name of Process>					
Purpose: <Concise description of the business reasons and objectives of the current process>					
Owner: <Identify the Process Owner>					
Suppliers	**Inputs**	**Outputs**	**Process**	**Customers**	**Systems**
• The internal or external roles and/or groups that provide required information and resources (inputs) to the performers of the process	• The information and/or resources provided to the performers of the process	• The product or service produced by the process	• The high level phases involved in the process i.e. Place & Confirm Order ⬇ Order Entry & Validation	• The internal or external roles and/or groups that receive outputs from the process	• The technology used to assist with the production of the outputs
Boundaries					
Start-point (Trigger): • The event(s) that occur(s) to commence the process			End-point: • The event(s) that occur(s) to signify the end of the process		
Includes: • List the characteristics of the transaction or scenario that are incorporated within this process • Some transaction types may be handled within this process, while other transaction types may be handled in a different process			Excludes: • List the characteristics of the transaction or scenario that are not incorporated within this process		

FIGURE 10-19 Process Profile Template

This helps reveal areas of dependency and gaps and can be used to understand the larger context in which a given process may operate. The exercise of completing the template has proven to bring items to the surface that otherwise would be left outside of the field of vision of the improvement project until they are discovered farther into design where it is more time consuming and sometimes more difficult to adjust a proposed solution. The process profile serves as a reference or definition for a process that helps others learn the process more quickly and more completely than simply reviewing a process map. The profile and the map are typically cross-referenced as supporting artifacts.

Process Value Analysis Template

A Process Value Analysis Template is used to identify process steps that add value and those that do not. It is also used to determine cycle time, bottlenecks, inefficiencies, defects, and/or waste. The template should be used to capture this information and enable unbiased assessment and comparison of value positions against proposed alternate process designs (see Figure 10-20). Value analysis helps to determine the overall health of a process by providing analytic information that determines if the process is meeting with expectations. This also helps determine its interoperability, its value in the context of the process ecosystem, the cost effective nature of the process, and the utility or usefulness of the process. The gaps discovered in the value analysis help identify areas for improvement.

To better understand the process:
- Review detailed process map
- Identify which of the process steps are value added and those that are non-value added
- Determine cycle time and identify bottlenecks
- Look for errors or inefficiencies that contribute to the defects

Process Step		1	2	3	4	5	6	7	8	9	10	Total Steps	Total Time	% of Total Time
Time (Hours)	Duration of process step													
Value-Added	In each step list the value added work - Customer recognizes the value in the step - Step changes the process													
Non Value-Added	In each step list the non-value added work - Customer does not recognize the value in the step - Step does not change the process **Allocate in each type below**													
Delay	This includes all idle time, such as waiting on up-stream operations and instructions.													
Set-Up	Setup time can include; switching between jobs; tracking down information; stalling until enough work piles up													
Inspection	Inspection is always considered to be a non-value added process. If it was done right in the first place there would be no need for inspection.													
Transporting	Moving materials in the process that can add to costs													
Rework	Correcting defects in the product is a major waste that includes material, labor, machine hours, or sorting													

FIGURE 10-20   Process Value Analysis template

Project Charter Template

A Project Charter Template is used to establish clarity around goals and objectives, scope, timelines, and resources. The template drives focus on what will be improved and what will not be improved (as shown in Figure 10-21).

Project Overview	
Project Name	Descriptive and concise name of the project
Department	The business unit being impacted
Requestor	Requestor of improvement project
Sponsor	Sponsor of improvement project (sponsor ensures alignment to strategic goals / Director level up)
Process Improvement Lead	Team member assigned to project
Requested Date (YYYY-MM-DD)	Original improvement request date
SLT Date (YYYY-MM-DD)	Senior Leadership Team delivery date
Start Date (YYYY-MM-DD)	Improvement project start date
Target Completion Date (incl. complexity level) (YYYY-MM-DD)	Estimated target completion date based on WBS milestones including complexity level chosen (high, medium, low)

Project Description	
Business Case	**Describe the reason why the organization should undertake this project.** Include; • The opportunity and the relationship to strategy • Explicitly why the project is being undertaken; is it to address opportunities or overcome challenges? • The value-add and rationale of the proposed initiative? Why is it important to do now? • What customers will benefit from the improvements • The ramifications if the project is not completed and implemented
Problem Statement	**Describe the problem or the current state as it is currently understood.** Include; • The pain or issue explicitly. What, where and when the problem occurs as precisely as possible • The current state, the desired state and the consequences of the current state process • The consequences of not addressing the problems of the current state process Do not; • Address more than one single problem • Assign or reference a cause • Assign blame • Offer or steer the reader towards a solution
Project Goals	**Describe what the project is focused on achieving as it is currently understood.** Defines and clarifies the goals the project is focused on achieving. The project is considered success only when these goals are met. Include; • Measurable targets. • Goals starting with a verb; eliminate defects; reduce cycle time; increase revenue; decrease costs

FIGURE 10-21   Project Charter Template

Estimated Benefits	Include; <ul><li>Quantified potential financial impacts linked to the business plan.</li><li>Estimates or qualitative statements that can be refined as the project proceeds and more data become available.</li><li>Both hard and soft benefits<ul><li>Hard benefit – Improvements that are real and measurable</li><li>Soft benefits – Improvements that do not immediately quantifiable; but provide benefits through efficiency, productivity, customer satisfaction and competitiveness</li></ul></li></ul>	
Project Scope	Area of Focus:	The projects primary area of focus; 'What process in what department in what location'?
	Includes:	Properly scoped projects are critical for success and optimal benefit. • What the project will include to meet the requirements of the project goals • What process(s), department(s), product types and regions are included in the project Do Not; • Include too much in the project scope • Overestimate the stakeholder's willingness to accept radical change within the specified timelines.
	Excludes:	• Out of scope excludes responsibilities, activities, deliverables or other areas that are not part of the project. • What process(s), department(s), product types and regions are excluded from the project
	Start Point:	Establish the start boundary of the process to be improved
	Stop Point:	Establish the end boundary of the process to be improved

Project Resources			
Team Members	**Function in the Business**	**Project Role**	**Time Dedicated (hours/week)**
Process Improvement Team Resource		• Provide mentoring and Process improvement tool expertise • Creates team meeting agenda and records action items • Chairs meetings and keeps focus by clarifying discussion points and testing for consensus • Oversee project schedule • Removes project	Estimated hours/week

FIGURE 10-21 *(Continued)*

		• roadblocks • Facilitate team communication and keeps the team on track • Maintains shared files	
Process Owner		• Provide input on critical decisions required • Approves and supports project • Approves changes in project scope and removes barriers • Provide resources to serve as team members and SMEs • Responsible for supporting implementation and improvement actions	Estimated hours/week
Subject Matter Experts (SME)		• As-needed position • Provides subject matter expertise to the project team • Can be assigned tasks within the team action plan • Delivers updates to team on status of actions steps, as needed • Acts as a change catalyst	Estimated hours/week
Project Team Members (other)		• Provides expertise and feedback to the projects team • Can be assigned tasks within the team action plan • Delivers updates to team on status of actions steps, as needed • Acts as a change catalyst	Estimated hours/week
Additional Support			

Project Milestones		
Milestone Phase	**Start Date**	**Completion Date**
Define		
Measure		
Analyze		
Improve		
Control		

FIGURE 10-21 (Continued)

The template captures the customer's requirements and the key attributes of the outputs. It succinctly articulates the problem to be solved using language and data as well as supporting information such as information from the 5 Whys. The project charter is also used as the artifact that documents the sanctioned approval and direction for a project team to execute against. In this regard, the performance improvement of a project can be revisited during the project as a health check to test for progress as well at project closure and at some time after project launch to assess the match of intent and results. Attributes within a charter include the project name, sponsors and participants, stakeholders, project leader/coach, and team members, the problem statement and objectives, the process profile (or SIPOC if running a six sigma project), the scope statement, and the overall work breakdown structure. A project charter may also include resourcing (people, technology, and money) needed to satisfy the goals of the project.

Project Closeout Communication Template

A Project Closeout Communication Template is used to support awareness and adoption of the change completion and identify the wins of the improvement. The template should be used at the end of a process improvement project when all the goals, objectives, and deliverables have been satisfied or when the project is ended due to cancelation or put on hold (see Figure 10-22). The communication should detail the formal acceptance of the solution and set the orderly process for exiting the project, including identifying the group responsible for ongoing support after the project is closed. The communication may also incorporate high-level statements about the success of the project, known issues, outstanding work for future consideration, acknowledgment of the effort, and results of the team supporting the project. It may also formally disperse the team and related resources for the project.

Project Plan Template

A Project Plan Template is used to identify and orchestrate all the various activities associated with the project to keep the project in control (as shown in Figure 10-23). The template should be used at the beginning of an improvement project and maintained throughout the lifecycle of the project. The template follows the Six Sigma DMAIC (Define, Measure, Analyze, Improve, and Control) model to articulate the work breakdown within each phase in order to keep the project on track. The DMAIC plan is a standard in the industry for running successful projects and encapsulates all key steps needed within a project. The template should be tailored to add additional steps and resources and to modify the timetable associated with the environment in which it is used. The ongoing progress of the project can be calculated based on the completion of time against the deliverable goal and the number of completed tasks/actions needed to meet with customer expectations.

Overview

In order to maintain consistency and ensure effective communications are sent on behalf of the Process Improvement team, there is a template created for project close-out communications. Please refer to the template and the example outlined in this document when sending communications after any project is completed.

Template

Hello,

Thank you for your participation in the "Enter Project Name" . I'm pleased to share our results.

"<Enter project accomplishments and align to initial goals>"

- •
- •
- •

"<Enter key project deliverables>"

If you have any questions, please feel free to contact me or the Process Improvement team at ProcessImprovement@yourcompany.com

Thank you,

"<Enter signature>"

Example

Good day,

Thank you for your participation in the Product Return As-Is Process Mapping Project. I'm pleased to share our results.

We were engaged by the Product Return team to create an As Is map in order for them to:

- • Highlight gaps, duplication of effort or activities that are not value added
- • Increase their understanding of how each department contributes to the end-to-end process
- • Identify potential process and system improvements

The Process Map and Process Profile have been completed and were added to the Process Ecosystem. The Map and associated Process Profile are also available at document repository. If you need a copy of the map printed on the plotter at any point, feel free to contact me or the Process Improvement team at ProcessImprovement@yourcompany.com.

Please let me know if you have any questions. Have a great weekend!

Sincerely,
Tristan Boutros

FIGURE 10-22 Project Closeout Communication Template

Timeline / Status

Timeline columns (all empty): 1-Sep | 8-Sep | 15-Sep | 22-Sep | 29-Sep | 6-Oct | 13-Oct | 20-Oct | 27-Oct | 3-Nov | 10-Nov | 17-Nov | 24-Nov | 1-Dec | 8-Dec | 15-Dec | 22-Dec | 29-Dec

No.	Action Steps / Deliverables	Start Date	Target Date	Revised Date	Status	Notes	Owner
	Milestone 1: Define						
1.01	Identify Stakeholders						
1.02	Write Project Charter						
1.03	Obtain approval through intake and governance to proceed with the improvement initiative						
1.04	Establish Project Plan						
1.05	Determine and map as-is processes						
1.06	Identify quick wins						
1.07	Understand the decision making landscape that impacts the project (eg. Create a stakeholder management plan)						
1.08	Create a plan to communicate project process						
1.09	Identify Voice of the Customer (VOC), translate into Critical to Quality (CTQ) metrics and define success measures						
1.10	Deliverable: documented VOC, CTQ and success measures						
1.11	Milestone: Tollgate Review with management						
1.12	Modify define phase deliverables based on Tollgate Review, if necessary						
1.13	Milestone: Define Phase Complete						
	Milestone 2: Measure						
2.01	Identify measures						
2.02	Create a data collection plan						
2.03	Engage business intelligence team to architect analytic solution						
2.04	Collect, interpret, describe and display data						
2.05	Deliverable: Report(s) and/or dashboards that measure(s) business performance						
2.06	Identify baseline performance from which improvements will be compared						
2.07	Milestone: Tollgate Review with management						
2.08	Modify Measure Phase deliverables based on tollgate review, if necessary						
2.09	Milestone: Measure Phase complete						
	Milestone 3: Analyze						
3.01	Perform Root Cause Analysis						
3.02	Conduct process analysis to determine non-value added steps						
3.03	Perform data and statistical analysis						
3.04	Deliverable: Root Cause analysis						
3.05	Milestone: Tollgate Review with management						
3.06	Modify Analyze Phase deliverables based on tollgate review, if necessary						
3.07	Milestone: Analyze Phase Complete						
	Milestone 4: Improve						
4.01	Identify and define potential solution(s)						
4.02	Determine and map potential to-be processes						
4.03	Engage Architecture team to determine the impact of the solutions on business, technology and data landscape, as needed						
4.04	Engage Technology Enablement team to determine the impact of the solution(s) on systems, as needed						
4.05	Conduct technology, risk, compliance and dependency assessment on proposed solution(s)						
4.06	Determine cost/benefit analysis of potential solution(s)						
4.07	Select the solution(s) to proceed with						
4.08	Inform Project Management Office (PMO) of proposed solution(s), as needed						
4.09	Prioritize solution(s) for iterative development and deployment						
4.10	Obtain management approval for solution selection and prioritization						
4.11	Transition implementation efforts to PMO, as needed						
4.12	Create and execute Process Change Management Plan						
4.13	Complete an implementation and training plan						
4.14	Update Process Library and job procedures						
4.15	Train Operators and management						
4.16	Deliverable: trained operators and management about the new procedures						
4.17	Test process and procedures for the solution						
4.18	Obtain change management approval for process changes						
4.19	Implement process changes						
4.20	Deliverable: implemented, validated, approved and documented solution(s) including process maps and procedures						
4.21	Milestone: Tollgate Review with management						
4.22	Modify Improve Phase deliverables based on tollgate review, if necessary						
	Milestone 5: Control						
5.01	Identify measures for process control						
5.02	Design process control dashboard						
5.03	Deliverable: report(s) and/or dashboard(s) to measure ongoing business performance						
5.04	Create a response plan for process issue resolution						
5.05	Deliverable: response plan for resolution of potential process issues						
5.06	Complete a process transfer plan for operators and management						
5.07	Communicate results to stakeholders and management						
5.08	Evaluate outstanding or new process improvement ideas discovered during the project						
5.09	Submit relevant process improvement ideas through process improvement intake						
5.10	Conduct post-project review for future considerations						
5.11	Milestone: Tollgate Review with management						
5.12	Modify Control Phase deliverables based on tollgate review, if necessary						
5.13	Close the project						
5.14	Milestone: control phase completed						

FIGURE 10-23 Project Plan Template

Project Status Report Template

A Project Status Report Template is used to provide updates to stakeholders, leaders, sponsors, and interested parties on the status and serves as the official record of communication about progress. The template captures accomplishments since the last update and the goals for the upcoming period and notes milestones and events (see Figure 10-24). The status

\<Project Name\> Project Status Report
\<Date of Message\>

Project Status: \<On Track, At Risk, Off Track\>

Start Date: *\<project start date\>*
End Date: *\<target project completion date\>*
Project Leader: *\<team member name\>*
Sponsor: *\<project sponsor name\>*

This Week's Accomplishments:

-
-
-

Next Week's Goals:

-
-
-

Upcoming Milestones:

-
-
-

If you have any questions or feedback, please feel free to contact *\<Name or Distribution List \>* or visit *\<website address\>* for more information.

\<Thank you\>,
\<team member\>

FIGURE 10-24   Project Status Report Template

Quick Win Opportunity	Easy To Implement	Reversible	Fast To Implement	Within Team's Control	Cheap To Implement
Process improvement techniques such as process mapping often uncover many easy and obvious improvement opportunities.	The improvement does not require significant planning or coordination.	The improvement is easily reversible to go back to the original process.	The improvement does not require a significant amount of time to implement.	The scope of the improvement is within the control of the team or sponsor of the project.	The improvement does not require a significant amount of capital and/or resources to implement.
1.					
2.					
3.					
4.					
5.					

FIGURE 10-25 Quick Win Identification Template

report is a powerful communication tool that can highlight important elements of the process improvement project that require attention and action. Status reports help keep engagement high, as well as drive alignment toward the process improvement project objectives.

Quick Win Identification

A Quick Win Identification is used to articulate the improvements that can be achieved in iterations and organizes the wins by benefit to the business/customer. Driving improvements often involves a series of realized benefits along a journey to a cumulative big win. Sorting out the elements that can be absorbed in atomic components, which independently add value and help the organization, can create confidence and help win support for further change acceptance. The template (as seen in Figure 10-25) helps to organize wins into segments that can be analyzed for the value each win offers as well as the level of effort or collateral impact associated with it. Quick wins vary in complexity but fundamentally resolve some sort of long-standing pain point the business is facing.

Response Plan Template

A Response Plan Template is used to document the method for how the process owner should respond to any out-of-control conditions that may occur in a process. The template aids in planning for contingencies and ensures consideration is given to the scenarios that are typical for many businesses so that the organization is best prepared for action (see Figure 10-26). Response plans include the mechanisms for who and how processes will be monitored, the parameters for when a response is warranted to events (e.g., a breach in a threshold), and what steps and activities are to be performed in such cases until the process is deemed in control. Response planning coordinates the staff during events and establishes a proactive state of operation. Response plans lower the time to recover during events and help mitigate risk of business disaster.

Measure	Action	Timing	Owner
The measure/metric that has shown the out-of-control condition	Specific action to be taken	Timing of action	Responsible to take action

Figure 10-26   Response Plan Template

Roles and Responsibilities Matrix

A Roles and Responsibilities Matrix is used to clarify accountability and responsibility for process steps organizationally. Assigning actions to specific roles sets the authoritative expectation for activity ownership and avoids duplication and confusion. The template identifies Performers, which indicates the step is performed by the function; Approvers, which indicates that approval is required by the function before the work can continue; Inputs, which indicates the input required to complete the step; Reviewers, which indicates functions that are required to review or be informed of the output; and Supporters, which indicates functions that support the completion of the step in the process. Figure 10-27 shows a

Roles & Responsibilities Matrix				
Process Steps	**Function A**	**Function B**	**Function C**	**Function D**
Step 1		I – Description of the inputs for this step	I – Description of the inputs for this step	
Step 2	P – Description of the step being performed	A – Description of the approvals required		
Step 3	A – Description of the approvals required		R – Description of the reviewers for this step	
Step 4		I – Description of the inputs for this step	P – Description of the step being performed	P – Description of the step being performed
Step 5	P – Description of the step being performed			

Steps are taken from the "As-Is" or "To-Be" process maps.
Functions are the departments depicted in the swim lanes from the "As-Is" or "To-Be" process maps.

P = Performer(s) = Indicates the step is performed by the function
A = Approval = Indicates that approval is required by the function before the work can continue.
I = Input = Indicates the input required to complete the step
R = Review = Indicates functions that are required to review or be informed of the output
S = Support = Indicates functions that support the completion of the step in the process

Figure 10-27   Roles and Responsibilities Matrix Template

Solution Description	Root Cause Addressed	CTQ Impact (H, M, L)	Implementation Effort (H, M, L)	Quick-Win (Yes, No)	Technology Required (Yes, No)
Description of each of the solutions proposed	How does the solution described address the root causes of the problem	Level of impact the solution will have on the CTQ	Level of effort of implementation given the available resources	Can this be classified as a quick win opportunity	Is a technology aspect required to implement to solution

Figure 10-28   Solutions Prioritization Matrix Template

standard Roles and Responsibilities Matrix Template. The matrix can be particularly useful in the context of understanding process responsibilities at the macro and micro levels. When using the matrix to attribute roles and responsibilities, the process for change management is made easier by providing clear information about the authority that controls a given process.

Solutions Prioritization Matrix

A Solutions Prioritization Matrix is used to describe how the various solution options compare to each other in meeting the goal of the improvement project. A sample template is shown in Figure 10-28. Presenting the solution in a matrix helps remove potential bias in the design of the solution by using standard comparators criteria. Each solution can be evaluated based on the value it offers, the level of effort it requires (both internally and by the customer), the dimension of benefit realization in time and investment, and resources required, such as technology. The solution prioritization should be customized to incorporate the criteria that are most meaningful in a given business situation. Criteria may not be equally weighted; in these cases, introduction of numerical values for each criterion should be considered.

Stakeholder Identification Template

A Stakeholder Identification Template is used to establish understanding of who a process impacts and/or who has a need to be engaged during potential process change consideration. Stakeholders are interested parties who have a stake or investment in the outcome of a process. This can include people who fund process work, operate the process, govern the process, receive the benefits of the outcome, or are dependent on or impacted in some way upstream or downstream of the process. Many people can be identified as a stakeholder; don't forget to incorporate the

Project Name: Descriptive and concise name of the project (should be identical to the Project Name in the Project Charter)			
Stakeholder Name	**Organizational Role**	**Project Role**	**Type (e.g., Internal, External, Supplier)**
The first and last name of the person or name of the group affected by the process or product the project team is working on	The job title of the person or group within the organization, whether internal or external	A description of the role the stakeholder will be performing on the project	A categorization of the stakeholder (Internal, External Customers or External Suppliers)

FIGURE 10-29 Stakeholder Identification Template

voice of the customer in some way in your stakeholder list (e.g., if the customer cannot be named directly, identify a proxy to act on their behalf and represent their needs and wants). Figure 10-29 outlines a standard Stakeholder Identification Template.

Stakeholder Management Plan

A Stakeholder Management Plan is used in tandem with the communication plan to make transparent the accountability and needs of the stakeholders for the improvement project and ensures that appropriate proactive engagement of stakeholders is maintained (see sample template in Figure 10-30). Some stakeholders are critical to implement a

Stakeholder	**Objective**	**Actions**	**Completion Date**	**Owners**
The name of the person the team needs to focus on	The purpose/desired outcome of the action	The method the proposed/desired outcome will be achieved (e.g., meeting)	The date the action will be completed	The person responsible for following through with the action

FIGURE 10-30 Stakeholder Management Plan Template

process improvement initiative, while others may be desirable but difficult to get time from, and still others need to be aware of the improvement but do not have a controlling interest in guiding the initiative. The Stakeholder Management plan creates the action plan associated with when different stakeholders will be engaged and what their roles are apart from others involved in the process, including what they are accountable for in the process.

Super System Map Template

Taken from the Rummler–Brache concept, a Super System Map Template is used to create the context of the organization and the business operation in which all processes operate. It aids in understanding the interdependencies between the organization and its environment. Drivers such as the labor market, the financial demands of shareholders, controls from legislation and regulatory bodies, and the influence pressure of competition are all consideration points in the super system map. Within the map, the concepts of performance expectation and performance management help to drive awareness and action toward tuning the organization toward high-quality output and sustainment practices. This template is the gold standard in helping employees and management understand their roles in the ecosystem of their business (see Figure 10-31).

Voice of the Customer–Critical to Quality Chart

A Voice of the Customer–Critical to Quality Chart is used to identify the expectations of the customer and identify the performance of a process toward that objective. This template reflects the qualitative and/or quantitative elements that drive customer satisfaction (see Figure 10-32). Sources include proactive interview, surveys, in-person visits and calls, market research, benchmarks, scorecards, and dashboards as well as reactive sources such as complaints, inbound calls, lost business, and increases in business demand. Regardless of the source, specific attributes are captured that identify the upper and lower limits of acceptable performance; these are used to design process improvement solutions. This template should be used during project definition, as it is a vital input into understanding the target improvement boundaries. What is unique about voice-of-the-customer information gathering is that it is an outside-in view from the customer that determines what performance impacts their satisfaction and identifies what the

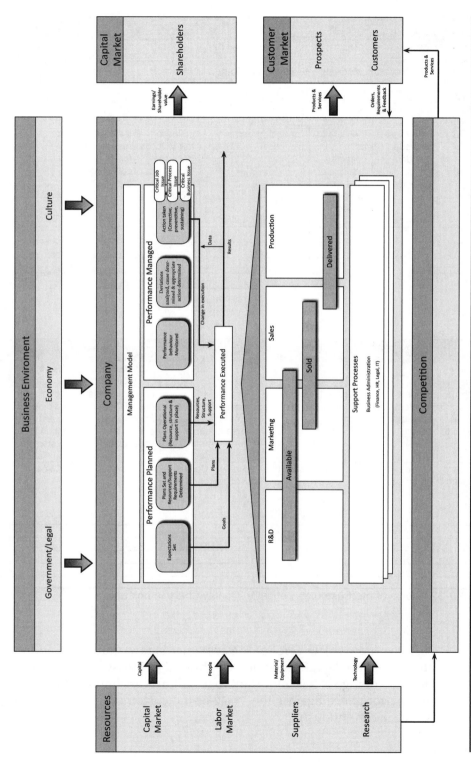

Figure 10-31 Super System Map Template

317

VOC Plan						
Customer Segments to Be Contacted for VOC	Individuals to Contact for VOC	Priority: High, Medium, Low	Information to be Gathered - Questions to be Asked	Data Collection Method	Project Team Owner	VOC Complete (✓)
Identify the groups that VOC needs to be collected from	Identifies the individuals or group that data will be collected from	Sets priority of the data to be collected in terms of importance to the project	Identifies the VOC questions to be used for each group or individual	The method for collecting the VOC (face to face interview, focus group, survey)	Who on the project team is responsible for collecting the VOC	Records when VOC is complete

CTQ Chart					
Customer Segment	Issue	Need	CTQ	Lower Spec Limit	Upper Spec Limit
The group that the issue has come from	What is the problem that the customer is seeing	What needs to be done	Translated customer need to a measureable requirement	Lower limit that the CTQ must fall within for customer satisfaction	Upper limit that the CTQ must fall within for customer satisfaction
Customers contacting the call center	Customers are on hold too long	Reduce hold time	Answer Calls <15 seconds	2 seconds	15 seconds

FIGURE 10-32 Voice of the Customer–Critical to Quality Chart Template

customer needs. The voice of the customer is raw, unfiltered, and pure and should not be reframed or paraphrased as intent can be misunderstood by doing so. It is important to revisit process improvement initiative performance against the voice-of-the-customer/critical-to-quality feedback continuously.

CHAPTER 11

Conclusion

The field of Process Improvement encompasses a wide variety of specialties including methodologies, frameworks, tools, techniques, aptitudes, and practices. As a unique and growing discipline, it has become necessary to establish standards of practice in the field. These standards have been established to identify the scope of competencies expected of Process Improvement practitioners, guide their professional training and development, and ensure an acceptable level of service to clients and organizations. Having established *The Process Improvement Handbook* as a body of knowledge, we aimed to

- Use an enterprise perspective to create a process ecosystem understanding
- Demonstrate the importance of end-to-end process improvement and the pitfalls of individual and isolated improvement efforts
- Outline the key steps in attaining higher process maturity and building a best-in-class Process Improvement organization
- Demonstrate linkage to business success, including success tips and tricks for the individual contributor, team, business division, supplier, and customer to extract high value with low investment
- Provide sample learning materials such as getting-started instructions, practical guides, real case examples, templates, and the process-oriented architecture road map
- Coach the reader on launching an Enterprise Process Architecture by providing a playbook for getting started and preparing the reader for rapid growth and increased visibility and demand for services
- Capitalize on practical agility principles such as iterative deployment to deliver fast results
- Demonstrate the value of process measures and their visibility to tuning organizational performance

The Process Improvement Handbook meets these goals. It has provided a useful and complete foundation, framework, and set of tools and principles for managing change and increasing organizational performance. The preceding chapters guided you through the appropriate steps to master the improvement process.

The *Process Improvement Context* set the stage for understanding the basic terminology and concepts of Process Improvement and demonstrated the value of several improvement methodologies while also presenting a common language for use in the industry. The *Process Maturity* chapter discussed levels of reliability, sustainability, and predictability used commonly in the industry to understand the baseline expectations for high performance. From this baseline the core concepts regarding the *Process-Oriented Architecture* (POA) and *Process Ecosystem* were deeply explored. Special attention was given to the overarching governance function in process orchestration, as so often improvement programs suffer from poorly managed governance activities. The functional walk-through in *Managing Process Improvements* served to review industry practices around understanding the improvement framework, environmental factors and organizational influences, an organizational profile, the role of leadership and strategic planning, and professional methodologies for process, performance, workforce, quality, and knowledge management. The Super System Map, Hoshin, Kaizen, Six Sigma, DMAIC (Define, Measure, Analyze, Improve, and Control), Plan-Do-Check Act, Total Quality Management, Rummler–Brache, and other approaches were discussed. The *Process Improvement Organization* and *Process Improvement Aptitudes* chapters examined the accountability, responsibility and skills, competencies, and capabilities needed to build and sustain a process-oriented enterprise. Competencies such as Analytical Thinking and Problem Solving, Culture of Measurement, Building Partnerships, Collaboration, Credibility, Flexibility, Initiative, Business Knowledge, Communication Skill, Interaction Skill, Sound Judgment, Resilience, Strategic, Situational Awareness, Building, and Recognizing and Retaining Talent were evaluated and described to ensure linkage to driving organizational transformation successfully. *Cases Examples* provided vivid accounts of the applicability of The Process Improvement Handbook in addressing improvement opportunities while striving to remind all practitioners that improvement work itself is a process that never ends.

The culmination of the material presented is only valuable when it is applied to real-life organizational need. We recommend that you build your own organizational process improvement framework based on the concepts, strategies, and knowledge presented in this text. *Templates* and *Instructions* that are ready to be tailored and used by your organization can be found in chapter 10, and for download on our companion website mhprofessional.com/pihandbook.

PART IV

Appendices

APPENDIX A

Process-Oriented Architecture Construct

Process - Oriented Architecture (POA)

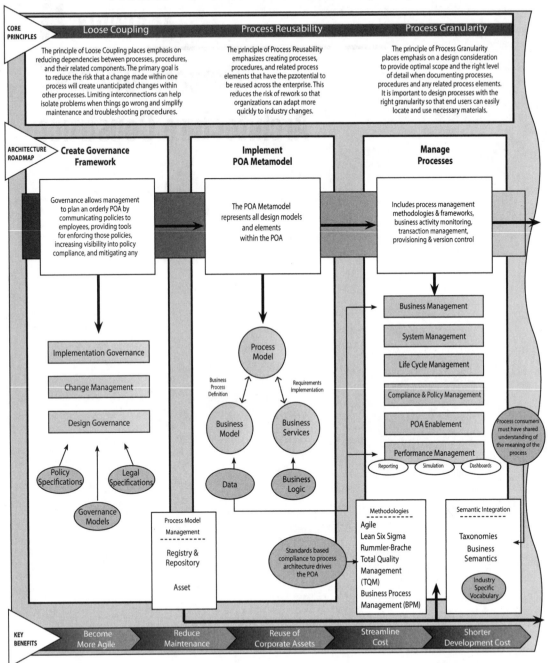

CORE PRINCIPLES

Loose Coupling

The principle of Loose Coupling places emphasis on reducing dependencies between processes, procedures, and their related components. The primary goal is to reduce the risk that a change made within one process will create unanticipated changes within other processes. Limiting interconnections can help isolate problems when things go wrong and simplify maintenance and troubleshooting procedures.

Process Reusability

The principle of Process Reusability emphasizes creating processes, procedures, and related process elements that have the pzzotential to be reused across the enterprise. This reduces the risk of rework so that organizations can adapt more quickly to industry changes.

Process Granularity

The principle of Process Granularity places emphasis on a design consideration to provide optimal scope and the right level of detail when documenting processes, procedures and any related process elements. It is important to design processes with the right granularity so that end users can easily locate and use necessary materials.

ARCHITECTURE ROADMAP

Create Governance Framework

Governance allows management to plan an orderly POA by communicating policies to employees, providing tools for enforcing those policies, increasing visibility into policy compliance, and mitigating any

Implementation Governance

Change Management

Design Governance

Policy Specifications

Legal Specifications

Governance Models

Process Model Management
- - - - - - - - - - - -
Registry & Repository

Asset

Implement POA Metamodel

The POA Metamodel represents all design models and elements within the POA

Process Model

Business Process Definition

Requirements Implementation

Business Model

Business Services

Data

Business Logic

Standards based compliance to process architecture drives the POA

Manage Processes

Includes process management methodologies & frameworks, business activity monitoring, transaction management, provisioning & version control

Business Management

System Management

Life Cycle Management

Compliance & Policy Management

POA Enablement

Performance Management

Reporting | Simulation | Dashboards

Process consumers must have shared understanding of the meaning of the process

Methodologies
- - - - - - - - - - -
Agile
Lean Six Sigma
Rummler-Brache
Total Quality Management (TQM)
Business Process Management (BPM)

Semantic Integration
- - - - - - - - - - -
Taxonomies
Business Semantics

Industry Specific Vocabulary

KEY BENEFITS

Become More Agile → Reduce Maintenance → Reuse of Corporate Assets → Streamline Cost → Shorter Development Cost

Process - Oriented Architecture (POA)

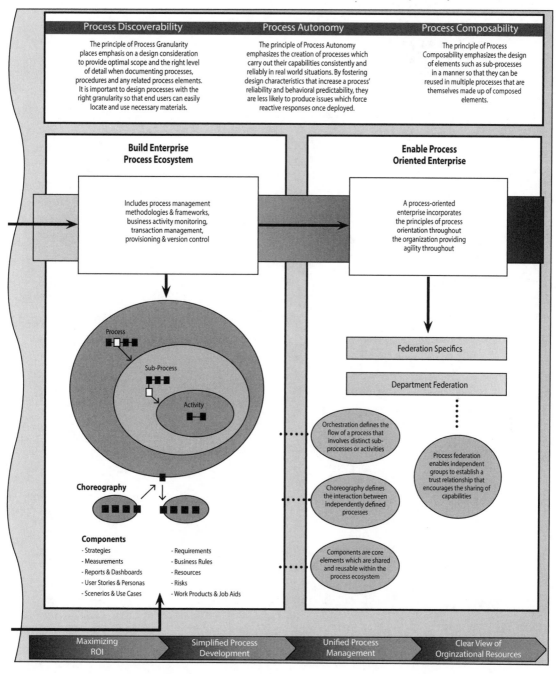

Process Discoverability

The principle of Process Granularity places emphasis on a design consideration to provide optimal scope and the right level of detail when documenting processes, procedures and any related process elements. It is important to design processes with the right granularity so that end users can easily locate and use necessary materials.

Process Autonomy

The principle of Process Autonomy emphasizes the creation of processes which carry out their capabilities consistently and reliably in real world situations. By fostering design characteristics that increase a process' reliability and behavioral predictability, they are less likely to produce issues which force reactive responses once deployed.

Process Composability

The principle of Process Composability emphasizes the design of elements such as sub-processes in a manner so that they can be reused in multiple processes that are themselves made up of composed elements.

Build Enterprise Process Ecosystem

Includes process management methodologies & frameworks, business activity monitoring, transaction management, provisioning & version control

Enable Process Oriented Enterprise

A process-oriented enterprise incorporates the principles of process orientation throughout the organization providing agility throughout

Process

Sub-Process

Activity

Choreography

Components
- Strategies
- Measurements
- Reports & Dashboards
- User Stories & Personas
- Scenerios & Use Cases
- Requirements
- Business Rules
- Resources
- Risks
- Work Products & Job Aids

Federation Specifics

Department Federation

Orchestration defines the flow of a process that involves distinct sub-processes or activities

Choreography defines the interaction between independently defined processes

Components are core elements which are shared and reusable within the process ecosystem

Process federation enables independent groups to establish a trust relationship that encourages the sharing of capabilities

Maximizing ROI

Simplified Process Development

Unified Process Management

Clear View of Orginzational Resources

APPENDIX B

Process Improvement Governance Structure

Process Improvement Governance

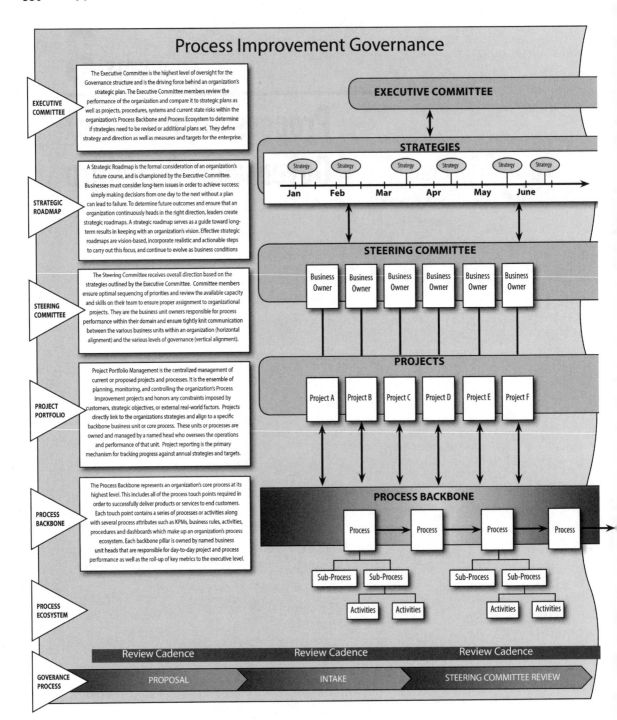

EXECUTIVE COMMITTEE

The Executive Committee is the highest level of oversight for the Governance structure and is the driving force behind an organization's strategic plan. The Executive Committee members review the performance of the organization and compare it to strategic plans as well as projects, procedures, systems and current state risks within the organization's Process Backbone and Process Ecosystem to determine if strategies need to be revised or additional plans set. They define strategy and direction as well as measures and targets for the enterprise.

STRATEGIC ROADMAP

A Strategic Roadmap is the formal consideration of an organization's future course, and is championed by the Executive Committee. Businesses must consider long-term issues in order to achieve success; simply making decisions from one day to the next without a plan can lead to failure. To determine future outcomes and ensure that an organization continuously heads in the right direction, leaders create strategic roadmaps. A strategic roadmap serves as a guide toward long-term results in keeping with an organization's vision. Effective strategic roadmaps are vision-based, incorporate realistic and actionable steps to carry out this focus, and continue to evolve as business conditions

STEERING COMMITTEE

The Steering Committee receives overall direction based on the strategies outlined by the Executive Committee. Committee members ensure optimal sequencing of priorities and review the available capacity and skills on their team to ensure proper assignment to organizational projects. They are the business unit owners responsible for process performance within their domain and ensure tightly knit communication between the various business units within an organization (horizontal alignment) and the various levels of governance (vertical alignment).

PROJECT PORTFOLIO

Project Portfolio Management is the centralized management of current or proposed projects and processes. It is the ensemble of planning, monitoring, and controlling the organization's Process Improvement projects and honors any constraints imposed by customers, strategic objectives, or external real-world factors. Projects directly link to the organizations strategies and align to a specific backbone business unit or core process. These units or processes are owned and managed by a named head who oversees the operations and performance of that unit. Project reporting is the primary mechanism for tracking progress against annual strategies and targets.

PROCESS BACKBONE

The Process Backbone represents an organization's core process at its highest level. This includes all of the process touch points required in order to successfully deliver products or services to end customers. Each touch point contains a series of processes or activities along with several process attributes such as KPMs, business rules, activities, procedures and dashboards which make up an organization's process ecosystem. Each backbone pillar is owned by named business unit heads that are responsible for day-to-day project and process performance as well as the roll-up of key metrics to the executive level.

PROCESS ECOSYSTEM

GOVERANCE PROCESS

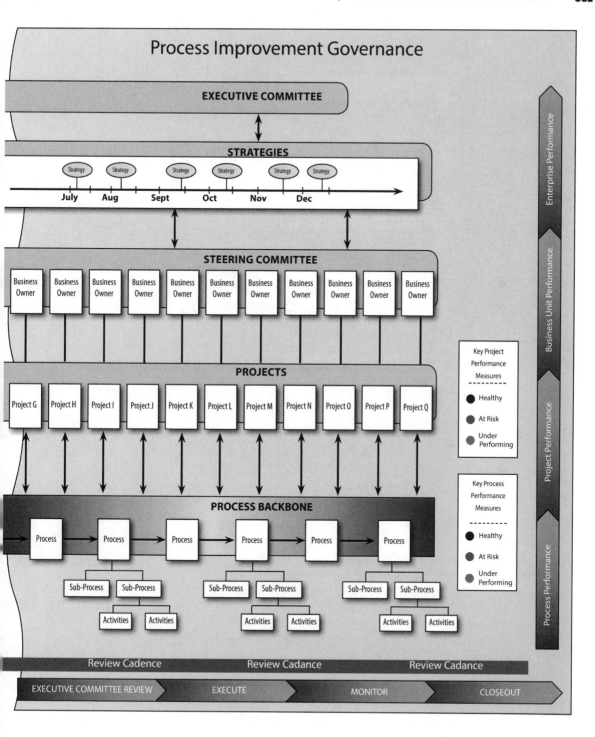

APPENDIX C
Acronyms

ADDDSA: Assess-Define-Develop-Deploy-Sustain-Adapt

CI: Continuous Improvement

CM: Change Management

CSF: Critical Success Factor

CTQ: Critical To Quality

DFLSS: Design-For-Lean-Six-Sigma

DMADV: Define-Measure-Analyze-Design-Verify

DMAIC: Define-Measure-Analyze-Improve-Control

EA: Enterprise Architecture

EM: Enterprise Modeling

EPM: Enterprise Project Management

FMEA: Failure Modes and Effects Analysis

HPS: Human Performance System

ISO: International Organization for Standardization

KM: Knowledge Management

KPI: Key Performance Indicator

KPM: Key Performance Measure

LSS: Lean Six Sigma

MSA: Measurement Systems Analysis

PA: Process Architecture

PDCA: Plan-Do-Check-Act

PDSA: Plan-Do-Study-Act

PI: Process Improvement

PIC: Process Improvement Coordinator

PIM: Process Improvement Manager

PIO: Process Improvement Organization

PMA: Process Maturity Assessment

PMM: Process Maturity Model

POA: Process-Oriented Architecture

RACI: Responsible, Accountable, Consulted, Informed

ROI: Return on Investment

SIPOC—Suppliers, Inputs, Process, Outputs, Customers

SMART: Specific, Measureable, Attainable, Relevant, Timely

SME: Subject Matter Expert

SOX: Sarbanes Oxley

SPI: Streamlined Process Improvement

SWOT: Strengths, Weaknesses, Opportunities, Threats

TOC: Theory of Constraints

TPS: Toyota Production System

TQM: Total Quality Management

VOC: Voice of Customer

APPENDIX D
Glossary

A3 Report Developed as a decision-making tool in the 1980s by Toyota Motor Corporation, it is a European paper size that is used to encompass various pieces of information on a single page to aid in understanding important information. The A3 document provides a structure and consistent format for communications and problem-solving methods. Many companies use the A3 report when planning improvements to processes.

5S A structured technique used to methodically achieve organization, standardization, and cleanliness in the workplace. The 5S pillars or practices are based on five Japanese words: Sort (Seiri), Set in Order (Seiton), Shine (Seiso), Standardize (Seiketsu), and Sustain (Shitsuke) and encourage workers to improve their working conditions and then reduce waste, eliminate unplanned downtime, and improve processes.

5 Whys A question-asking technique used to explore the root cause of a particular defect or problem.

Affinity Diagram Also called an Affinity Chart, is a graphical tool used to help organize ideas generated in brainstorming or problem-solving meetings.

Agility The ability of a business to adapt and change course both rapidly and cost efficiently in response to changes in the organizational environment.

Analytical Thinking A process that emphasizes breaking down complex problems into single and manageable components.

Assess-Define-Develop-Deploy-Sustain-Adapt (ADDDSA) The five steps taken during a Rummler–Brache project where a new process is being created or an existing process is being improved.

Balanced Scorecard (BSC) A Balanced Scorecard is a structured dashboard that can be used by managers to keep track of the execution of activities by the staff within their control and to monitor the consequences arising from these actions.

Benchmarking The process of comparing one's business processes and performance metrics with those that demonstrate superior performance in their industry or from other industries.

Bowling Chart   A dashboard used to track performance (plan versus actual) on Strategic Planning objectives.

Brainstorming   A technique used to generate a large number of creative ideas within a group or team of people.

Building Partnerships   The ability to build mutually beneficial business relationships that foster improved business outcomes.

Business Knowledge   Having the business acumen needed to drive strategy, evaluate, and improve performance.

Business Rules   Statements that define or control aspects of a business and its processes.

Catchball   A technique used to manage group dialogue and develop ideas, tasks, metrics, and strategies to support the accomplishment of an organization's strategic goals.

Change Management   The process of transitioning individuals, teams, and organizations from a current state mentality or way of operating to a desired future state.

Charter   A mandate to clarify the various responsibilities associated with the centralized and coordinated management of Process Improvement efforts and projects across an enterprise.

Check Sheet   An organized way of collecting and structuring data so that decisions can be made based on facts, rather than anecdotal evidence.

Coaching   The process of enabling stakeholders to grow and succeed by providing feedback, instruction, and encouragement in order to assist them in discovering solutions on their own.

Collaboration   Embracing the ability to work harmoniously with others in the business environment toward a common purpose.

Communication   The process of conveying information through the exchange of speech, thoughts, messages, text, visuals, signals, or behaviors.

Competencies   Also referred to as Key Competencies, are those behaviors and capabilities an individual has that make him or her stand out as superior performers.

Compliance Management   Often referred to as Policy Management, is an organization's approach to mitigating risk and maintaining adherence to applicable legal and regulatory requirements.

Continuous Assessment   The act of continually monitoring and assessing processes and operations using performance metrics and acting on those results to improve performance.

Controls Management or department controls can be defined as a systematic effort by business management to institute predetermined standards, plans, or objectives in order to determine whether performance is in line with these standards and presumably in order to take any remedial action required to see that human and other corporate resources are being used in the most effective and efficient way possible.

Credibility Having the ability to establish the trust needed to help organizations move through transformation stages more smoothly.

Corporate Culture The beliefs and behaviors that determine how a company's employees and management interact and handle outside business transactions.

Decision Making The process of choosing what to do by considering the possible consequences of different choices.

Define, Measure, Analyze, Design, Verify (DMADV) The five steps taken during a Lean Six Sigma project where a new process needs to be created.

Define, Measure, Analyze, Improve, Control (DMAIC) The five steps taken during a Lean Six Sigma project where an existing process needs to be improved.

Design Governance An organization's process for monitoring the activities associated with modeling current and desired processes and architecting and proposing process changes.

Enterprise Architecture (EA) Also known as Business Architecture, the act of transforming an organization's vision and strategy into effective enterprise change by creating, improving, and communicating the requirements, principles, and models that describe an organization's current operations and enable its evolution.

Enterprise Modeling A technique used to diagrammatically architect the structure, processes, activities, information, people, goals, and other resources of an organization.

Environmental Factors Also known as Organizational Influences, the internal and external factors that surround or influence an organization, a department, or a project.

Facilitation The process of taking a group through learning or change in a way that encourages all members of the group to participate.

Fishbone Diagram Also known as a Cause and Effect diagram or Ishikawa Diagram, is a tool used to visually display the potential causes of a specific problem or event.

Flexibility The ability to adapt to change, shift focus and resources, and manage through the change.

Force Field Analysis A technique used to identify forces that may help or hinder achieving a change or improvement within an organization.

Framework   The foundation upon which an organization creates, delivers, and captures value. It is a structured guide intended to steer an organization toward building particular practices or philosophies into its culture and operations.

Functional Management   An organizational management structure where departments are grouped by areas of specialty (e.g., sales, marketing, manufacturing).

Governance   The activity of directing and controlling an organization and its activities with the purpose of defining expectations, granting power, or verifying performance.

Human Performance System (HPS)   A model that outlines several variables that can influence the behavior of a person in an organizational work system.

Implementation Governance   The practice of managing process changes and additions through the process improvement lifecycle to ensure modifications are appropriate and sustainable.

Initiative   The ability to take immediate action regarding a challenge, obstacle, or opportunity while thinking ahead to address future challenges or opportunities.

Job Aides   Devices or tools (such as instruction cards, memory joggers, or wall charts) that allow an individual to quickly access the information he or she needs to perform a task.

Just Do It   The most basic concept of Process Improvement, it describes the immediate implementation of a specified process change. It is used primarily when a problem with a process has been identified, the solution is known and understood, and very little effort is required to implement the change.

Kaizen   A Japanese term for the gradual approach to increasing quality and reducing waste through small but recurrent improvements that involve all workers in an organization.

Kaizen Event   Highly concentrated team-oriented efforts used to quickly improve the performance of a process. Typical timelines are less than five days.

Key Performance Indicators (KPI)   Measures of progress toward achieving a stated Key Performance Measure.

Key Performance Measures (KPM)   A set of indicator metrics represented at key points along a process or group of processes that radiate performance about a process.

Leadership   The act of motivating and inspiring a department or group of employees to act toward achieving a common goal.

Lean Six Sigma (LSS)   A systematic method for improving the operational performance of an organization by eliminating variability and waste within its processes.

Loose Coupling   A Process-Oriented Architecture principle that focuses on reducing dependencies between processes, procedures, and their related components.

Management Processes   Business processes that govern the operations of an enterprise or organization.

Manifesto   An article that sets forth the principles, core values, and/or goals of an organization or work product.

Metrics   The defined measures of an organization that assist with facilitating the quantification of a particular characteristic.

Monitoring   The act of continually measuring processes to watch for irregularities in performance.

Negotiation   A process that encompasses a discussion between two or more individuals who seek to find a solution to a problem that meets both of their needs and interests acceptably.

Operational Governance   An organization's structure for ensuring processes are behaving correctly.

Operational Processes   Also known as primary or core processes, are business processes that form the primary objective of the enterprise and subsequently create the primary value stream.

Organizational Charts   An organizational chart is a diagram that shows the structure of an organization and the relationships and relative ranks of its parts and positions/jobs.

Organizational Profile   A high-level overview of an organization that addresses the organization's operating environment, its key relationships, its competitive environment and strategic context, and its approach to performance improvement.

Pareto Diagram   Also known as Pareto Chart, is a bar chart that is typically used to prioritize competing or conflicting problems or issues so that resources are allocated to the problems that offer the greatest potential for improvement. This is done by showing their relative frequency or size in a bar graph. A representation of data in the form of a ranked bar chart that shows the frequency of occurrence of items in descending order.

Performance Assessment   A service provided by Process Improvement organizations to evaluate performance, delivery of services, and quality of products provided to consumers, as well as the performance of any human resources involved in process execution.

Performance Management   The process by which companies ensure alignment between their employees and both company and department goals and priorities.

Performance Management Cycle   A method used to evaluate desired behavior against actual performance.

Personas   Detailed descriptions or depictions of the customers and users of a process. Personas are based on real people and real data that humanize and capture key attributes about archetypal process customers.

Plan-Do-Check-Act (PDCA)   A four-step problem model used in Process Improvement activities. Also known as the Deming Wheel, Deming Cycle, or Plan-Do-Study-Act Process.

Poka-Yoke   A Japanese term that means mistake-proofing a process, procedure or activity. It can be considered as any mechanism in a process that helps an operator avoid mistakes or eliminates product defects by preventing, correcting, or drawing attention to human errors as they occur.

Policies   Basic principles and/or guidelines formulated in order to direct and limit the actions of an organization.

Procedure   An outline of the specific instructions that describe how to perform the steps or activities in a Process.

Process Architect   The individual responsible for architecting and designing an organization's processes.

Process Autonomy   A Process-Oriented Architecture principle that emphasizes the creation of processes that carry out their capabilities consistently and reliably in real-world situations.

Process Backbone   An organization's core process at its highest level. This includes all process touch points required in order to successfully deliver products or services to end customers.

Process Choreography   The act of sequencing several processes to run in connection with one another.

Process Composability   A Process-Oriented Architecture principle that emphasizes the design of process elements such as subprocesses in a manner that they can be reused in other organizational processes.

Process Dashboard   A user interface comprised of graphical information (charts, gauges, and other visual indicators) that displays current or historical trends of an organization's various Process Performance Measures.

Process Discoverability   A Process-Oriented Architecture principle that emphasizes making processes discoverable by adding interpretable Meta Data to increase process reuse and decrease the chance of developing processes that overlap in function.

Process Documentation   The document artifacts that describe the process, inputs, controls, outputs, and mechanisms associated with a process.

Process Ecosystem   A modeling concept that describes how all processes are interconnected and driving toward business success in the context of their internal and external environment.

Process Federation   The act of consolidating and aggregating process-related information across multiple sources into a single view.

Process Granularity   A Process-Oriented Architecture principle that places emphasis on a design consideration to provide the right level of detail to process users when documenting processes, procedures, and any related process elements.

Process Improvement   Often called Continuous Improvement, is the ongoing effort of an organization to improve its processes, services, or products.

Process Improvement Aptitudes Dictionary   A document that strives to outline the underlying characteristics that define the patterns of behavior required for Process Improvement professionals to deliver superior performance in their roles. There are 10 skills and 10 competencies that are considered essential for ensuring the success of individuals involved in Process Improvement efforts. When combined with various Process Improvement methods and techniques, can collectively deliver process excellence.

Process Improvement Coordinator   The individual who is generally responsible for maintaining any procedural and administrative aspects of the Process Improvement Organization and serves as the designated point of contact for all Process Improvement issues and questions.

Process Improvement Handbook   A collection of instructions that are intended to provide ready reference in a formal text and that describe the tools, techniques, methods, processes, and practices needed for successful Process Improvement delivery.

Process Improvement Manager   The individual responsible for developing the capability of an organization by teaching Process Improvement skills and managing any Process Improvement projects or related endeavors.

Process Improvement Organization (PIO)   A group assigned with various responsibilities associated with the centralized and coordinated management of Process Improvement efforts and projects across an enterprise.

Process Management   The ensemble of activities related to planning, engineering, improving, and monitoring an organization's processes in order to sustain organizational performance.

Process Map   A step-by-step pictorial sequence of a process that illustrates its activities, inputs, outputs, departments, subprocesses, and various other business attributes.

Process Mapping The act of visually describing the flow of activities in a process and outlining the sequence and interactions that make up an individual process, from beginning to end.

Process Maturity Assessment A diagnostic tool used to appraise and characterize an organization's processes.

Process Maturity Model A set of structured levels that describe how well the behaviors, practices, and processes of an organization are reliably and sustainably producing required outcomes.

Process Monitoring The act of monitoring and analyzing a processes performance, including its relevance, effectiveness, and efficiency.

Process Monitoring Plan A document that outlines how process performance will be continuously monitored, who will be notified if there is a problem, how that will happen, and what response will be required.

Process Orchestration The coordination and arrangement of multiple processes or subprocesses exposed as a single aggregateprocess.

Process-Oriented Architecture (POA) The philosophical approach to process architecture, development, and management.

Process Owner The individuals within an organization who are accountable and responsible for the management and improvement of the organization's defined processes and related process components.

Process Reusability A Process-Oriented Architecture principle that emphasizes creating processes, procedures, and related process elements that have the potential to be reused across an organization.

Process Simulation The imitation of how a real-world process or series of processes will perform over a specified period time.

Process Training The act of providing and/or receiving education in the formal discipline of process improvement (e.g., *The Process Improvement Handbook*, Lean Six Sigma, Rummler-Brache).

Processes A series of activities or subprocesses taken to achieve a specific outcome.

Project An individual or collaborative initiative designed to determine the cause of a business or operational issue, define and analyze current performance of activities related to the issue, and implement improvements that rectify the issue and ensure that appropriate operational transition and monitoring occur.

Project Intake The process used to ensure all Process Improvement service requests are properly assessed for prioritization, resourcing, and execution.

Project Portfolio Management The centralized management of current or proposed projects and processes.

Project Proposal   The articulation, typically in documented business case format, of a new project or process change request aligned with an organization's strategic plan. A proposal may come about as a result of new information derived from organizational performance, market opportunities and/or changes in the business environment.

Resilience   The ability to remain resilient and persistent in pursuing goals despite obstacles and setbacks.

Resources   The components required to carry out the activities of a given process or series of processes.

Responsible-Accountable-Consulted-Informed (RACI)   The participation of various roles in executing a business process or Process Improvement project. A tool used to clarify roles and responsibilities.

Retrospective   A meeting that Process Improvement project teams hold upon release of a process change into the organization or at the conclusion of a Process Improvement project to reflect on what activities went well and what activities went poorly and to discuss actions for improvement going forward.

Risk   Any factor that may potentially interfere with or impact the successful execution of an organization's processes, projects, or operations.

Root Cause Analysis (RCA)   A method of problem solving that tries to identify the root causes of process fluctuations or problems that cause operating events that affect performance.

Rummler–Brache   A systematic approach to business process change. A step-by-step set of instructions on how to make changes to the way work gets done across an organization.

Sarbanes-Oxley Act (SOX)   An act passed by US Congress in 2002 to protect investors from the possibility of fraudulent accounting activities by corporations.

Scatter Diagram   Also called a Scatter plot or X–Y Graph, is a tool for analyzing relationships between two variables. One variable is plotted on the horizontal (X) axis and the other is plotted on the vertical (Y) axis.

SIPOC Diagram   A tool used by a Process Improvement team to identify all relevant elements of a process before Process Improvement work begins.

Situational Awareness   The ability to make sensible decisions based on an accurate understanding of one's current environment.

Sound Judgment   The ability and willingness to be unbiased and objective when making business decisions based upon consideration of various alternatives.

Specific-Measurable-Attainable-Relevant-Timely (SMART)   A mnemonic used to guide organizational leaders toward setting specific objectives over and against more general goals.

Sponsor   Also known as Project Sponsor, the person who has a central role in Process Improvement initiatives and is responsible for the funding and overall direction of a Process Improvement project.

Spot Awards   An award that recognize individuals on an impromptu basis for their special effort to their department or an improvement initiative.

Steering Committee   A cross-functional executive group that sets agendas, strategic objectives, schedules of business, and overall parameters. Typically comprised of various department managers and process owners, these are the individuals who evaluate organizational performance as well as track progress of projects set forth to improve performance across an organization.

Stop-List   A list of processes, terms, or business activities that should be ignored, bypassed, or retired by a particular organization.

Strategic Goals   Planned objectives that an organization strives to achieve.

Strategic Implementation   The ability to link strategic concepts to daily work.

Strategic Planning   The act of defining an organization's strategies or business direction and making decisions regarding the allocation of resources to pursue those strategies, including capital and people resources.

Strategic Road Map   A strategic road map is a visual plan that offers goals and strategies for the future of a business, organization or group. Just like a physical map helps a driver navigate the route to his or her destination, a strategic road map aids an organization in following a path to its goals.

Subject Matter Expert (SME)   A person with in-depth knowledge to share about a topic being presented or discussed or someone who is a specialist in a specific knowledge area.

Subprocess   A series of activities that accomplish a significant portion or stage of a higher order or parent process.

Super System Map   A picture of an organization's relationships with its business environment, customers, suppliers, and competition that helps it understand, analyze, improve, and manage those relationships.

Supporting Processes   Business processes that support the core operational processes of an organization.

Systems Management   The management of the information technology systems within the various processes in an enterprise.

Team Achievement Awards   Awards that recognize teams within or across departments that complete an important process improvement or milestone.

Template   A template is a master copy of a publication used as a starting point for various business efforts including project management, process improvement, and various other activities. It may be as simple as a blank document in the desired size and orientation or as elaborate as a nearly complete design with placeholder text, fonts, and graphics that need only a small amount of customization.

Time Management The process of planning and exercising control over the amount of time spent on specific activities in order to increase effectiveness, efficiency, or productivity.

Training A process aimed at bettering the performance of individuals and groups in organizational settings.

Tree Diagram Also known as a Systematic Diagram, a technique used to break down broad categories of content into finer levels of detail.

User Stories Brief descriptions of process change requests or enhancements told from the perspective of the person who desires the new capability, usually a process operator or customer.

Voice of Customer (VOC) The needs or requirements, both stated and unstated, of an organization's customers.

Warranty Period A predetermined length of time during which performance after project implementation is measured against the expected target.

White Space As described by Rummler–Brache, the area between the boxes in an organization chart or the area between the different functions or departments within an organization. Rules are often vague, authority fuzzy, and strategy unclear, resulting in misunderstandings and delays.

Work Instructions A subset of procedures that are typically written to describe how to do something for a single role or activity within a process.

Workforce Engagement An organization's system for developing and assessing the engagement of its workforce, with the aim of enabling and encouraging all members to contribute effectively and to the best of their ability.

Workforce Management An organization's system for developing and assessing the engagement of its workforce, with the aim of enabling and encouraging all members to contribute effectively and to the best of their ability.

APPENDIX E
Illustrations and Figures

Section I - Chapter 01:

- Figure 1-1 Process Improvement Manifesto

Section I - Chapter 02:

- Figure 2-1 Sample Policy
- Figure 2-2 Sample Process Map
- Figure 2-3 Sample Procedure
- Figure 2-4 Process Hierarchy
- Figure 2-5 Distinguishing Characteristics of a Policy, Process & Procedure
- Figure 2-6 Sample Functional Organization Design
- Figure 2-7 Sample Process Focused Organization Design
- Figure 2-8 Leadership Strategies for Process Improvement

Section II - Chapter 03:

- Figure 3-1 Process Maturity Model

Section II - Chapter 04:

- Figure 4-1 Process-Oriented Architecture Construct
- Figure 4-2 Process-Oriented Architecture Guiding Principles
- Figure 4-3 Process Ecosystem Overview

Section II - Chapter 05:

- Figure 5-1 Sample Process Ecosystem Groupings
- Figure 5-2 Sample Persona of a Process Customer
- Figure 5-3 Common Business Rules Attributes
- Figure 5-4 Process Ecosystem Concept
- Figure 5-5 Process Ecosystem Structure

APPENDIX F

References

Acuity Institute. "Lean Six Sigma Black Belts—Define-Measure-Analyze-Improve-Control Phases." 9th ed. Print.

Bailey, Ian. "A Simple Guide to Enterprise Architecture."2006. http://www .cioindex.com/enterprise_architecture/ArticleId/75005/A-Simple-Guide-to-Enterprise-Architecture.aspx.

Boisvert, Lisa. "Strategic Planning Using Hoshin Kanri Hoshin Kanri." Hoshin Kanri for Strategic Planning. May 2012. http://www.goalqpc .com/HoshinKanriStrategicPlanning.cfm.

Brassard, Michael and Ritter, Diane. *The Memory Jogger 2: Tools for Continuous Improvement and Effective Planning*. 2nd ed. Goal/QPC, 2010.

Business Process Maturity Model (BPMM): What, Why and How, http://www. bptrends.com/publicationfiles/02-07-COL-BPMMWhatWhyHow-CurtisAlden-Final.pdf.Accessed January 7, 2013.

Continuous Improvement—ASQ—Learn About Quality—Overview, PDCA Cycle. "ASQ: The Global Voice of Quality." Web. 10 Feb. 2011. <http://asq.org/learn-about-quality/continuous-improvement/overview/overview.html>.

Deming, William Edwards. "Out of the Crisis." Cambridge, MA: Massachusetts Institute of Technology, Center for Advanced Engineering Study. February1982.

Global Standard, PMI. *A Guide to the Project Management Body of Knowledge (PMBOK Guide)*. 4th ed. Newtown Square, PA: Project Management Institute, 2010.

Harrington, H. James. *Streamlined Process Improvement*. 1st ed. New York: McGraw-Hill Professional, 2011.

"HR for Employers | Government of Canada—Continuous Learning."*HR for Employers | Government of Canada—Home Page*. 30 Mar. 2009. Web. 10 Feb. 2011. <http://www.hrmanagement.gc.ca/gol/hrmanagement/site.nsf/eng/hr11570.html>.

Jacobs, F. Robert, William L. Berry, D. Clay Whybark, and Thomas E. Vollman. *Manufacturing Planning and Control for Supply Chain Management*. New York: McGraw-Hill/Irwin, 2011.

Kemsley, Sandy. "Business Process Modeling." Business Process Management (BPM) Resource Center. http://www.tibco.com/solutions/bpm. Accessed January 29, 2013.

Mcloed, Graem. "The Difference Between Process Architecture and Process Modeling." 'Inspired!' March 18, 2009. http://grahammcleod.typepad .com/inspired_knowledge_and_in/2009/03/the-difference-between-process-architecture-and-process-modeling.html.

MCTS-What is Hoshin Kanri?—Management Coaching & Training, http://www.mcts.com/Hoshin-Kanri.htm. Accessed February 1, 2013.

National Institute of Standards and Technology (NIST). "Criteria for Performance Excellence." *2009–2010.* Print.

Organizational Maturity Assessment—Hewlett-Packard. http://h10076. www1.hp.com/education/om.pdf.Accessed February 7, 2013.

Pritchett Company. "Improving Performance." Rummler-Brache | Improving Performance. Web. 20 Mar. 2011. <http://www.rummlerbrache.com/ home>.

Pyzdek, Thomas, and Paul A. Keller. *The Six Sigma Handbook.* 3rd ed. New York: McGraw-Hill Professional, 2009.

Rummler, Geary A., and Alan P. Brache. *Improving Performance: How to Manage the White Space on the Organization Chart.* San Francisco: Jossey-Bass Publishers, 1990.

Schwaber, Ken. *Agile Project Management with Scrum.* Redmond, WA: Microsoft Press, 2004.

Weill, Peter, and Jeanne W. Ross. *IT Governance: How Top Performers Manage IT Decision Rights for Superior Results.* Boston: Harvard Business School Press, 2004.

Index

Note: Page numbers referencing figures are followed by an "*f*."

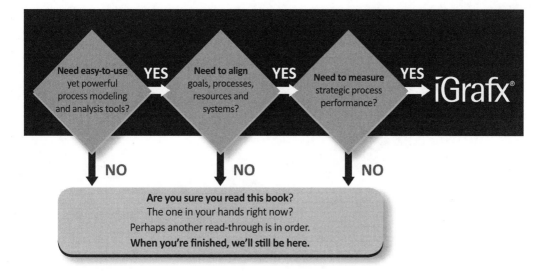